# THE SUPERIOR
## NORTH SHORE

▲

# THOMAS F. WATERS

# THE SUPERIOR

# NORTH SHORE

▲

Illustrations by Carol Yonker Waters

▲

Maps by the author

▲

University of Minnesota Press ▲ Minneapolis

The John K. Fesler Memorial Fund and David R. Fesler
provided assistance in the publication of this volume,
for which the
University of Minnesota Press is grateful.

Published by the University of Minnesota Press
2037 University Avenue Southeast, Minneapolis, MN 55414.
Published simultaneously in Canada
by Fitzhenry & Whiteside Limited, Markham.
Printed in the United States of America.
Designed by Gale L. Houdek.

**Library of Congress Cataloging-in-Publication Data**

Waters, Thomas F.
  The Superior North Shore.

  Bibliography: p.
  1. Natural resources—Minnesota. 2. Natural
resources—Ontario. I. Title
HC107.M6W38 1987   333.7'09776'7   87-13585
ISBN O-8166-1613-2

The University of Minnesota
is an equal-opportunity
educator and employer.

To
John B. Moyle and
Lloyd L. Smith, Jr.
scholars of the North Shore
colleagues
friends

# The Call of the Wild Shore*

*Have you ever stood on the savage shore,*
*With the storm upon your eye,*
*And watched the Herring Gull come wheeling*
*Down a gray November sky?*

> *Have you tramped the trails of the voyageurs?*
> *Have you paddled their rushing stream?*
> *Have you heard the laugh of the loon at night*
> *'Neath a star-sky all agleam?*

*Have you peered down in the lava rifts,*
*Hell-rent and heaven-spurned?*
*Where now the amber liquid glides,*
*But once the glaciers churned?*

> *Some say it's a land that's been tortured*
> *By hellfire hotter than sun.*
> *Some say that it's cold and they love it;*
> *Some want to go back—and I'm one!*

*You say you've gazed on the granite hills,*
*That grace the eastern shore?*
*And felt the hush of darkened valleys*
*Thrill you to the core?*

> *But hear you the howl of Windigo?*
> *On a bleak and storm-spent shore*
> *On a night so dark the snow is black*
> *And the storms of December roar?*

*But then see the muted sunrise*
*Climb an ice-limned lava strand,*
*And watch the purple fog-squall fade—*
*Stark beauty the eye can't stand.*

> *So now come straddle the lava strand*
> *Where the breakers rip and roar,*
> *And then come boast of this northern coast,*
> *Of Lake Soo-peer-ee-OR!*

*With my great respect to Robert Service, bard of the North.

# *Contents*

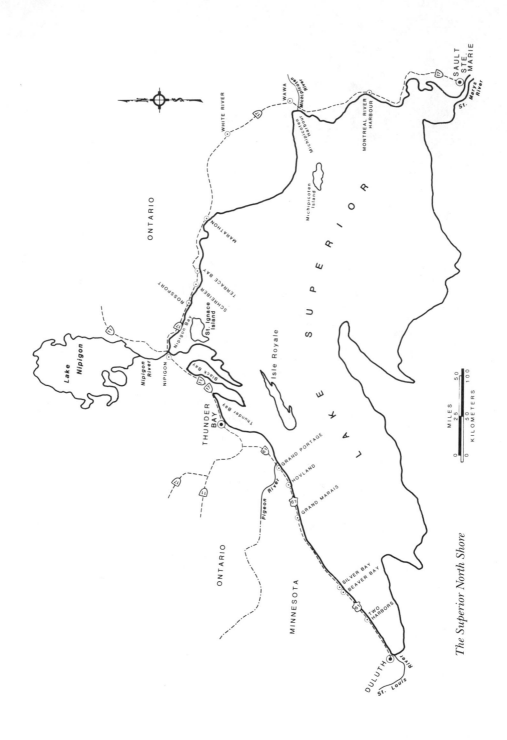

*The Superior North Shore*

# *Preface*

This book relates the events of the creation and succession of natural resources in the North Shore region of Lake Superior. It also tells the tales of human use and exploitation of these resources—as well as some of our attempts to preserve their natural character.

Accounts of geologic formation, human colonization, the adventure of exploration, the wrenching of livelihoods from this wild and northernmost of Great Lakes shores, the successes and failures of attempts to garner fortunes from natural resources in a new land—these are exciting tales. Exciting also are our attempts to correct past mistakes, as well as to meet the challenges ahead of us.

My purpose in this book is to bring to the reader knowledge, which in turn may bring about a sense of intimacy and wonder. Intimacy with landform, river flow, and seascape—and the dynamic forces produced by climate and fertility—should, I hope, produce a sense of proprietorship, and thus stewardship.

I define the North Shore and take as my geographic scope the strands, forests, and lakes and rivers between the headwater stream entering Lake Superior at the western end—the St. Louis River—and the lake's outlet at the eastern end—the St. Marys River. North of this line, I include roughly the Lake Superior drainage, but environments and events on lands, lakes, and rivers outside this drainage basin, relevant in some way to the North Shore, are also included. Thus defined, the North Shore is enormously rich in history and resources, and the study of it has become an immense and unfinished task.

Many tales remain untold. The aboriginal Ojibway possessed no written history, and the archaeological record is skimpy. Neither was the wilderness conducive to journal writing in the European style. An inhospitable, infertile region, the rugged coastline and its hinterlands combined to present a great obstacle—and yet passage—to exploration and commerce. And while some travelers kept diaries, white Europeans who first penetrated the turbulent waters and menacing spruce forests probably were more intent upon survival than they were upon record keeping for posterity.

And yet, as the wilderness was removed and the land more densely populated, technologies and communication were improved. Natural resources were discovered and tapped, and political organization was developed. More journals and personal records were kept. Scientific explorations were begun.

Many books, personal histories, and scholarly publications have thus accumulated over the last 200 years—roughly corresponding to the time when citizens of both Canada and the United States undertook permanent settlement and commenced an unprecedented harvest of the region's natural resources. One of my objectives in this volume is to stimulate the reader toward continued study of the North Shore through its rich literature.

Some of these publications treat in greater depth the natural resource subjects included in this volume; there is a wealth of further reading on the fur trade, the logging era, mining for iron and the rarer metals, and fish and wildlife.

Other publications, not directly concerned with natural resources, deal more with human history on the North Shore: the early map makers, shipping and shipwrecks, railroad building, early settlements and modern cities. In a later section of this book, in the front of the Bibliography, I have included some suggestions for the reader interested in pursuing many aspects of the history and character of the Superior North Shore.

Throughout the region's history of exploration, scientists and scholars left a wealth of new knowledge—beginning with naturalist Louis Agassiz, then surveyor Robert Bell, geologists C. R. Van Hise and C. K. Leith, ecologist William Cooper, ichthyologist John Dymond, fisheries biologists Lloyd Smith and John Moyle, wildlife biologist Durward Allen, and a literal army of following scholars and students who continue to explore the infinite mysteries of changing natural ecosystems. The limnology of Lake Superior has been studied intensively; even so, deepwater scientific research continues today in a new dimension with a submersible named *Sea Link II*, which in the summer of 1985 permitted the first direct, human observation of the lake's icy but biologically active bottom, far below sea level.

My treatment of natural resource subjects may sometimes seem illogical or inequitable to the reader.

For example, much more space is devoted to *fish* than to other animal groups. Good reasons exist, I believe, for this apparent unfairness.

First, among the several major kinds of natural resources in the North Shore region, the fishery of Lake Superior's nearshore waters stands out as the one having the greatest impact upon human culture along the coast. Primitive peoples and historic Indians alike were vitally dependent upon this bounty from the inland sea; their life-style was often built around fish and fishing. Later white cultures—from the fur trade to modern tourist industry—depended heavily upon fish for food and income.

Second, the effect of the sea lamprey depredation upon the Great Lakes' fish community was regarded by many as the most significant ecological calamity, greatest in its scope, in North America's recent history. Along the Superior North Shore, the disruption caused by the lamprey to fish populations, fishermen's livelihoods, and life-styles of all inhabitants was at once deep and dramatic. The tale is complex; the telling of it has been a fascinating but large task.

Third, I cannot deny some personal bias in devoting an inordinate proportion of the book to fish and fishing—for which I ask some charitable forbearance.

Another inequality might be inferred from the practice of capitalizing the common names of birds but not those of other animals. Professionals in the field of ornithology have adopted this convention for good reason: to make it clear which species are being referred to. For example, there are many species of white-throated sparrows, but only one White-throated Sparrow.

Within the greater ecosystem of the North shore are two special subunits, nearly microcosms by themselves: Isle Royale and Lake Nipigon. Both are large, and so are they both *insular*, in an ecological sense: Isle Royale is surrounded by water, Lake Nipigon by land. Within Isle Royale's land mass are smaller lakes; and within Nipigon's broad waters are smaller islands. Each has a natural and human history distinct from—though related to—the larger cosm of the North Shore, and thus each is accorded a chapter to itself.

My greatest hope is that the pages in this volume will stir the reader to explore personally the rugged coast of the North Shore and its backcountry. For here are to be found at first hand the cascading rivers and quiet estuaries that carried the canoes of explorers and voyageurs; here are the ranges of hills and forests that once attracted the prospector and timber cruiser; here are many islands, ranging from the great archipelagoes of Isle Royale and Nipigon Bay, still attracting the curious scientist, down to wooded islets and tiny wave-battered rocks holding the fascinating botanical oddity of disjunct Arctic flowers.

Personal sensations and perspectives of the traveler on foot must remain intimate, for they will vary with each individual who receives in different ways the impressions of beauty or repulsion, of solace or fear, or a moving experience of wildness. The richest perceptions of the senses can only be attained on the spot—in the sound of crashing surf or waterfall, the smell of cedar in a moist gorge or jack pine on a sunny upland plain, the visual reception of a bursting sunrise over tranquil swells or a seascape under storm-darkened skies.

To help you get there, take advantage of a profusion of excellent guides, listed in the Bibliography. Maps, lists of campgrounds, guides to fishing, hunting, and hiking, brochures on parks and forests, navigation charts, and boat access guides—all are available from state, provincial, and federal agencies by the packsackful.

My own perceptions of the land north of Superior began nearly fifty years ago, as a teenager accompanying

my father and his fishing pals to a log cabin on the shore of
Speckled Trout Lake, District of Algoma, Ontario, in pursuit of
brook trout. We caught the Algoma Central at the Sault, the train
huffing and puffing for four hours to cover exactly 100 miles,
around shining blue lakes and fractured granite hills, over
wooden-trestled chasms and whitewater rivers. My recollection
was that the tracks were so winding we could look out the window
almost any time and see the caboose going by in the opposite
direction.

It was a lonely feeling, at first, when the train pulled away
and left us with our pile of baggage, food, and tackle at the side
of the tracks. The silence was palpable. My other main perception
at that point was one of wonder at the dark tangle of spruce and
fir on all sides, mixed with the anticipation of catching those big,
red, speckled beauties in the days ahead.

The first trip of the year, in May, was a time of cold fingers
and unexpected snows, but a rebirthing of old fishing excite-
ments; June and July were horrors of black flies, and for the most
part we waited them out at home. But September—what glory!—
crimson maples and golden birches, frosty mornings and crisp af-
ternoons when the big specks were redder and wilder than ever
as they crashed the surface for our flies. We looked forward to
crisp, fried fillets and a blazing fireplace every evening. To the
west only a few miles, we knew, was Lake Superior, an inland sea
remote and huge, unknown.

The Second World War, with its gasoline rations and barrage
balloons over the Sault locks, put a crimp in our visits to Speckled
Trout Lake, and for me a complete hiatus for a time. My father
and mother, fishing on a mid-August day in 1945 from a rowboat
in the little bay next to the railroad tracks, received their first no-
tice of war's end with the joyous shrieking of the steam whistle of
the Algoma Central, hurtling north along the eastern shore to
spread the news. And soon I was to come back for more of those
speckled beauties.

Many years and college lectures later, I visited the North
Shore to receive perceptions of another sort—on the lava-
armored coast of Minnesota's northwest shore. I was there in an
academic role now, to attempt the solution of scientific mystery:
the functioning of biological systems in the rivers and streams
that, in such profusion, tumbled into Lake Superior. It was a sys-

tematic quest this time, deliberate and analytic. But those ventures brought new perceptions of this unusual region's geologic and natural history—its wild shores and forests and rivers.

In a few more years, an automobile road was completed around the great lake. This development, of course, dispelled some of the remoteness, but it also opened the previously denied great adventure of a complete circumnavigation around the inland sea. To drive to Speckled Trout Lake—well, almost—approaching from the wild northwest, was a marvel.

I marveled, too, at the incongruity of the political boundary. Here, one of the North Shore's most beautiful streams, the Pigeon River, separated two nations.

But the North Shore was not so fractured in the earlier centuries of the opening of the Northwest. The rich lake trout fisheries and bountiful fur-bearing populations did not recognize any such boundary. Nor did the intrepid French explorers and British fur traders have to stop at the Pigeon River to tell where they were born and how much money they had spent on one side or the other.

Geological forces created the North Shore of Lake Superior eons ago as a unique ecosystem—a region where earth features and biological organisms occurred with a high degree of organization, not simply in random chaos—and not in two portions.

It is in this sense that I have prepared this book. My purpose, in part, has been to break down that border. For me, broiled fresh lake trout taste as delicious on both sides; the spectacle of a bull moose thrashing toward shore through floating lily pads is just as grand in the District of Algoma as in Cook County; the pleading cry of a Herring Gull over a stormy Superior strand is spine tingling anywhere along the thousand-mile coast.

My self-imposed assignment of integrating two nations across an artificial but officious border was not undertaken without apprehension. With no roots in Canadian culture, was I not presumptuous in the extreme to think I could be accepted and assisted—a necessity to carry out my task—in asssembling material from the Canadian side?

I needn't have worried. It is a delight to report that one of my greatest satisfactions in the entire process of preparing this book was the enthusiastic assistance I received from new Canadian friends. For me, the Pigeon River is a boundary no more.

The assistance of many persons in both countries was essential for assembling material and providing field opportunities. This group embraced nearly the entire spectrum of persons working or interested in the North Shore region: professors and former students, librarians, commercial fishermen, agency biologists, anglers and hunters and loggers, museum curators, birders, local history buffs. All around the shore, requests for assistance were never refused. Since the principal emphasis was on natural resources, I relied heavily on both the Minnesota Department of Natural Resources and Ontario's Ministry of Natural Resources; their partnership was vital.

I owe much to the staff of my own Department of Fisheries and Wildlife at the University of Minnesota—secretaries, colleagues, students—for uncountable courtesies that smoothed my way on numerous occasions over many years.

Assisting in many ways, often repeatedly, by providing materials, consulting, writing and searching on my behalf, answering my inquiries, were a host of persons who always leaped to give me help. Those with whom I had the closest association were: David Bean, Ontario Ministry of Natural Resources, Nipigon; Ray Bonenberg and Carol Dersch, Ontario Ministry of Natural Resources, Lake Superior Provincial Park; Peter J. Colby, Ontario Ministry of Natural Resources, Thunder Bay; Francesca J. Cuthbert, Department of Fisheries and Wildlife, University of Minnesota, St. Paul; William J. Dalton, Department of Biology, Lakehead University, Thunder Bay; E. D. Frey, Ontario Ministry of Natural Resources, Wawa; Claude E. Garton, Department of Biology, Lakehead University, Thunder Bay; D. W. Hodgins and Karen Tierney, Pukaskwa National Park, Marathon, Ontario; Manfred M. Kehlenbeck, Department of Geology, Lakehead University, Thunder Bay; Conrad P. Krueger, angler, Minnetonka, Minnesota; L. M. Lein, Nipigon Historical Museum; Walter T. Momot, Department of Biology, Lakehead University, Thunder Bay; Stanley S. Sivertson, commercial fisherman, Duluth; Stanley L. Smith, U.S. Fish and Wildlife Service, Twin Cities, Minnesota; Barry J. Snider, Ontario Ministry of Natural Resources, Terrace Bay; Jeffrey N. Stone, College of Forestry, University of Minnesota, St. Paul; James Storland, Minnesota Department of Natural Resources, Grand Marais; Evan R. Thomas, Ontario Ministry of Natural Resources, Wawa.

I visited and corresponded with a number of libraries around the shore. I imposed the most on the following librarians, whose help in locating materials, sometimes in little-known and far-off places, was critical: Katherine Chiang, Barbara Kautz, and Susan Stegmeir, Entomology, Fisheries, and Wildlife Library, University of Minnesota, St. Paul; Jacqueline Baker, Old Fort William Library, Thunder Bay; and Marjorie King, The Chancellor Paterson Library, Lakehead University, Thunder Bay.

Several trips in the field, to forest, rivers, and the great lake, were both delightful and productive in the perceptions I received of special North Shore resources. Those days of discovery and fellowship, some of my most cherished North Shore experiences, were generously provided by new and some old friends: Ross Chessell and Tony Damiani, Ontario Ministry of Natural Resources, Nipigon; John deLaittre, Beaver Bay Club, Beaver Bay, Minnesota; Thomas W. Eckel, commercial fisherman, Grand Marais, Minnesota; Robert Hamilton, Ontario Ministry of Natural Resources, Thunder Bay; William K. Robinson, Ontario Ministry of Natural Resources, Terrace Bay; and John R. Spurrier, Minnesota Department of Natural Resources, Duluth.

Preparation of a book of this kind entails much expense. I am deeply grateful for the generosity of a number of persons and foundations, intimately concerned for the history and resources of the North Shore. Charles H. Bell and John deLaittre assisted greatly in helping me to raise funds and followed closely my progress with the book. The following provided financial support: The Quetico Superior Foundation, principally through the efforts of R. Newell Searle; the David Winton Bell Foundation; Sewell D. Andrews, Jr.; Atherton Bean; C. M. Case, Jr.; Thomas M. Crosby; Virginia and Leroy R. Genaw; Louis W. Hill, Jr.; Clinton Morrison; and Phillip W. Pillsbury.

Adding enormously to the accuracy and completeness of factual material in the entire book were technical reviews made by specialists in each of the fields given a separate chapter. Many gave much valued further technical help. I am most grateful for their willingness to expend time and critical effort toward the production of an authoritative account of North Shore resources. This group comprised Clifford E. and Isabel Ahlgren, Quetico-Superior Wilderness Research Center, Ely, Minnesota; Jeff Allen, Ontario Ministry of Natural Resources, Chapleau; Charles R.

Burrows, Minnesota Department of Natural Resources, retired, Minneapolis; Joel Cooper, Ministry of Natural Resources, Wawa; K. G. Fenwick, Ontario Ministry of Natural Resources, Thunder Bay; Janet C. Green, Minnesota Ornithologists' Union, Duluth; John C. Green, Department of Geology, University of Minnesota, Duluth; Robert Hamilton, Ontario Ministry of Natural Resources, Thunder Bay; Richard L. Hassinger, Minnesota Department of Natural Resources, St. Paul; Peter A. Jordan, Department of Fisheries and Wildlife, University of Minnesota, St. Paul; Murray W. Lankester, Department of Biology, Lakehead University, Thunder Bay; Ralph W. Marsden, Department of Geology, Emeritus, University of Minnesota, Duluth; Jean Morrison, Old Fort William Library, Thunder Bay; A. L. Robinson, Ontario Ministry of Natural Resources, Wawa; John P. Ryder, Department of Biology, Lakehead University, Thunder Bay; George R. Spangler, Department of Fisheries and Wildlife, University of Minnesota, St. Paul; Lorne Townes, Ontario Ministry of Natural Resources, and others of his staff at Nipigon: Blake Beange, Rick Borecky, Ross Chessell, Richard Raper; James C. Underhill, Department of Ecology and Behavioral Biology, University of Minnesota, Minneapolis; Alan R. Woolworth, Minnesota Historical Society, St. Paul; Herbert E. Wright, Jr., Department of Geology and Geophysics, University of Minnesota, Minneapolis; and J. V. Wright, Archaeological Survey of Canada, Ottawa.

My greatest pleasure in preparing this book was in the companionship and sharing of adventure with my wife, Carol, during my sabbatical year on the North Shore. We hiked, waded, canoed, and rock-scrambled together in many beautiful places. And in our trailer parked along lakes, rivers, and Lake Superior coastlines, all around the shore, we outlined much of what this book was to say and how it would be illustrated. There were many occasions when she stayed faithfully at work sketching, while I, alas, went fishing.

T. F. W.

# THE SUPERIOR

## NORTH SHORE

# La Marge Sauvage

A billion years ago.

Distant in earth's history, toward the end of the ancient Precambrian Era, middle North America was rent asunder in a great rift.

This breach in the earth's Precambrian crust extended more than 1,000 miles in length, curving through the northern part of the region that was eventually to contain the Laurentide Great Lakes. It is the northern edge of this crescent that today forms the rocky, rugged North Shore of Lake Superior.

The breach—the Midcontinent Rift System—varied in width, and it was not always continuous. About 50 to 100 miles wide, it stopped short of separating the continental land mass into two parts and creating a major ocean between.

But at the same time, because of the tremendous tensional wrench caused by this separation of the earth's crust, the region was subjected to one of the most active periods of volcanic activity that North America had yet undergone. The creation of the rift and its associated volcanism did not occur with one gigantic crack; rather, volcanic activity was to continue over a period of perhaps 100 million years.

From deep in the earth, an immense volume of molten rock, or magma, worked its way upward toward the surface. Some of the magma solidified in great underground masses; some poured out as lava into the primordial atmosphere over yet older rock. Time after time, lava flowed from fiery vomitories in the rending earth to crystallize in marvelous variety.

Lava spread out into huge "lava lakes" to accumulate into plateaus miles thick, only to founder into a great sink. It was a major feature in the earth's surface, 350 miles long, 250 miles wide—this primordial basin of Lake Superior. Eventually, in another billion years, it would contain the greatest expanse of fresh water in the present-day world. The rifting and associated volcanism were the last of volcanic earth-shaping events in the North Shore region.

The Precambrian Era left a foundation of ancient rock crust under the whole of northeastern and central North America, virtually set apart from the rest of the earth's history. For, during all of remaining geologic time—when elsewhere in the world great plates of the earth's crust floated about on the molten interior and collided, mountains rose and were eroded, and vast seas advanced and receded—and life evolved—this basement of Precambrian rock lay almost unchanged.

Winds and rivers eroded the less resistant of the older rocks, and only a few thousand years ago glaciers from the north scoured depressions and left their sediments. Nevertheless, the great rock base remains today essentially as it was after the upheaval of the rift formation—and the North Shore as it was a billion years ago.


The French called it *le lac supérieur*—not for the superior qualities we recognize today, but simply because in their grand sweep of exploration and adventure through the Great Lakes region three centuries ago, they found Lake Superior the *uppermost*, the final headwater.

Prehistoric tribes had their own name for such an immense body of water. Spelled out in many variations by different European chroniclers, the Ojibway's *Kitchi Gami* ("great lake,") appears to be the most accurate. Its headwaters occur close to those of the *Mississippi* the Ojibway's ("great river"). But unlike the name *Mississippi*, *Kitchi Gami* was not adopted by recorded history, but rather the Europeans' *Superior*.

4

Into the wilderness of the Northwest in the seventeenth century only two major water routes were available to European explorers attempting to penetrate the region from the Atlantic side of the continent. The northern wilderness took in a million square miles of forests, streams, and sprawling, spruce-lined lakes—most of that which was to become western Canada. The vast area seemed a tangled maze of waterways—but one easily penetrable by the travel mode of the native Indians, the birchbark canoe. The waterways beckoned with a great attraction—the beaver—and the Europeans were drawn irresistibly by this immense region of land and water rich with fur, highly regarded in Europe.

The first route into the region was by way of Hudson Bay, extending saltwater navigation westward virtually into the center of North America, rapidly claimed by the English and their Hudson's Bay Company.

The other route comprised the St. Lawrence River and the chain of Great Lakes, extending westward from New France colonies around the wide St. Lawrence estuary to the uppermost of these freshwater seas. And here the mostly flatwater route ended. Three relatively small rivers—the Kaministiquia, Pigeon, and St. Louis—emptied their waters from upland origins, their headwater sources leading farther yet into the Northwest. But the lower waters of all three rivers presented major obstacles to upstream travel—waterfalls and cataracts, tumbling over hard ledges and through rocky defiles to the North Shore of Lake Superior.

The three rivers, at different times and for canoe brigades of different nationalities, together constituted a major segment of the route into the Northwest. All three necessitated arduous portages around lower turbulent waters; the upper passages, by portage and paddle, ranged first upstream and then down. All led to Rainy Lake and Lake of the Woods, then by the Winnipeg River to Lake Winnipeg, where the canoe routes split into the myriad of waterways that wove through western and northern Canada. For nearly two centuries these three rivers poured a wealth of furs, probably numbering into the millions, from the Northwest to the North Shore, requiring enormous effort and many lives of French-Canadian voyageurs.

Yet this incredible enterprise left little permanent mark on the Northwest wilderness and the North Shore. Old lodges and stockades decayed into the forest mold; beaten trails and portages were reclaimed by ground vegetation and new forest growth; and the beaver, brought close to extirpation, returned to its former abundance, as wild animal populations always will do when their habitat remains intact and suitable. The fur traders' and voyageurs' legacy now consists of a few reconstructions or restorations of their major establishments, seasonal camps, and portage paths now dedicated in national, state, and provincial parks and monuments, some French and Indian names for lakes and streams—and a lusty history rich in adventure.

The North Shore and its hinterlands were many things to different peoples. Certainly to the early white Europeans, it was a gateway: to lands beyond the roots of old mountains where lay the discovery of rivers, new mountains, oceans— possibly a passage to the Orient. To Jesuit missionaries, many souls to be saved. To French and English traders, the route to untold riches in fur.

Later, the exposed lavas and granites of this northern coastline offered first copper, later gold and silver, to the hard-rock prospector. And then the *red-gold* of ore from all around the coast produced the iron that was to be manufactured into the hardware of battle, employed on fields of conflict during two world wars.

To the lumberman, the pine and spruce forests offered the riches of *green-gold*, an evergreen source of lumber to build the cities of North America and of pulpwood to create the newspapers of the world.

To the prehistoric Ojibway, the North Shore held its symbols of mystery and power. Not the easiest to locate, on Hat Point, near Grand Portage, the ancient *witch tree* still grows—a 400-year-old white cedar, gnarled, time-and weatherbeaten. Defiantly clinging to eroded and lichen-covered rock only a few feet from Lake Superior's waves, it was revered and feared by the Indians who brought gifts to appease its spirit or gave it wide berth in their canoes. The witch tree also caught the eye of explorer Sieur de la Vérendrye, 250 years ago.

Today, near the boundary of two modern nations, the witch

*The Witch Tree (courtesy of Susan Genaw)*

tree stands as a symbol of North Shore endurance, unyielding to its trials and its centuries.


The boreal forest, with its dark and shadowed spruce cathedrals, is a spooky place. Even on the clearest of days, overhead only the most intricate tracery of blue sky shows beyond branches of black spruce, and underfoot are spongy *Sphagnum* and other mosses, never touched by rays of the sun. But to native Americans, descendants of Ice Age cultures, the boreal forest was the source of staple food and clothing: the ancient woodland caribou, also a descendant of an Ice Age. The boreal forest left its ghostly mark also on native culture; for when winter blizzards howl in the black spruce tops far above the drifted snow, it is not the wind but rather the shriek of giant cannibals of the winter woods, Manitous of Ojibway legendry.

This was *la marge sauvage*—the wild shore—a coast whose tumultuous tributaries beckoned the adventurous with discovery, the entrepreneur with mineral and fiber, the sports enthusiast with new species of game and fighting fish, with sheer cliffs to climb and wild rivers to run.

It remains a shore of strange beauty, when in a winter dawn the sun rushes upon a frozen strand of jagged rock, and stranger yet on a misty estuary when the loon throws its chilling cry across darkening waters.

It was a savage land, too, of sudden death for the French voyageur, whose sense of thrill overcame common sense when he tried to run a river rapid, ripped by high waters of springtime— or whose frail birchbark craft, overladen with the season's take of beaver pelts, was caught by a sudden storm in Superior's wide waters, too far out from harbor.

And it was also a shore unique on the continent in its tectonic history that attracted the attention of later scholars in their attempts to unravel Earth's geologic record through billions of years.


Plain statistics can be boring, but those for Lake Superior are at least impressive. Its area is the largest of the world's freshwater lakes: 31,500 square miles (81,900 square kilometers). Its volume: 2,860 cubic miles (11,920 cubic kilometers), 10 percent of the earth's liquid fresh water. (Not the world's single greatest volume of fresh water, however—that distinction belonging to Lake Baykal, in the U.S.S.R., at 5,450 cubic miles.) Superior's elevation: 602 feet (183 meters) above sea level, its waters dropping from that height through its outlet in the St. Marys River, Lakes Huron and Erie (Lake Michigan is a dead-end cul-de-sac), over Niagara Falls to Lake Ontario, and finally down the St. Lawrence River to the Atlantic Ocean. The deepest part of Lake Superior's bottom lies more than 1,330 feet (405 meters) beneath the surface of its waters, well below sea level.

Lake Superior's average (and usual) temperature: a chilly forty-three degrees Fahrenheit (eight degrees Celsius), although some shallow bays, especially on the east side, are a bit warmer. The constancy of water temperature affects local climate—especially when winds blow off the lake. Winters are milder, summers cooler, along the shore. Springtime may be delayed, and autumn colors may come late, up to a week. The reason for such a significant effect, of course, lies in the great volume of water, requiring enormous amounts of energy to change from any temperature. Freezing is uncommon, occurring only in bays and shoals; icing of the entire lake has occurred only once of record, in 1979.

The shape of the lake today is still like a giant crescent—the legacy of the ancient Precambrian rift—with Thunder Bay and Nipigon at its top, or "Lakehead." Irregularities and islands along the shorelines produce a shape that bears a striking resemblance to the head of a wolf, with the jaws (separated by Michigan's Keweenaw Peninsula) slightly open, with Isle Royale as its eye, and Duluth at its nose.

Lake Superior's published statistics offer its length at 350 miles (560 kilometers) and maximum width at 160 miles (260 kilometers). But if you traveled by boat from Sault Ste. Marie, Ontario, to Duluth, Minnesota, down the center of the lake (equidistant from both shores), your course would entail a journey of over 400 miles (650 kilometers). Shipping distance, between Sault Ste. Marie and Duluth, is given at 376 miles; the sunrise requires about one-half hour for the trip in the same direction.

But what about the northern coastline, the North Shore—how long? Minnesota's portion is remarkably straight and is commonly, if somewhat variably, given at about 200 miles (300 kilometers); Ontario's portion, with many more bays, estuaries, and peninsulas, is a little over 900 miles (1,475 kilometers), for a wandering total of 1,100 miles, or 1,775 kilometers. By car, the North Shore Drive and the Trans-Canada Highway, from Duluth to Sault Ste. Marie, you would drive 650 miles (1,000 kilometers). If you were in a hurry and wanted to cut some corners (but not many), you could board a light plane and fly the full length of the coast, crossing peninsulas and bays, and you would fly 600 miles (nearly 1,000 kilometers).

For the French-Canadian voyageurs in their historic, annual trips by canoe and paddle—traversing minor embayments and river estuaries, avoiding islands and turbulent shoals, making daily landings—it was an incredible journey of probably 1,000 miles (1,500 kilometers), from Sault Ste. Marie to Fond du Lac, the fountainhead of Lake Superior.

But hike the water's edge—assuming if you could scuttle across the facades of vertical lava cliffs and wade the mouths of turbulent rivers—and it would be twice again as far.

Some hike.

Some North Shore.

# ONE

# *Genesis by Fire*

T he age of the earth, as geologists reckon time with
modern radiometric tech-
niques, is estimated at ap-
proximately 4.5 billion years. The first billion will always remain
somewhat of a mystery, for this was the time when the planet was
formed by rapid accretion of space-borne particles, ranging in
size from dust to asteroids. Continuing bombardment, radioac-
tive energy, and increasing density caused the temperature of the
interior to rise greatly, and internal matter reached the molten
stage.

Nothing remains from this period for us to see and analyze
now, and so our knowledge of what happened in the North Shore
region during that remote time must remain inferential.

But when the crust later cooled and thickened to a surface of
water and rock, and an atmosphere developed, a visible and durable record was begun. It is through the study of the rocks—
their chemical composition, their crystalline structure, their
placement and orientation, and the alignment of magnetic particles—that we can today read the history of our planet.

We are fortunate indeed that the area north of Lake Superior is one of few regions where we can read some of the earth's
earliest record in its rocks.

Geologists have divided the history of the earth into several major eras. Each of these included events significantly different from those in other eras. The first era—the Precambrian—is by far the longest (nearly 90 percent of the total!), and it is the one with which we are mostly concerned on the North Shore. In this embryonic time of our planet, the surface was formed into a solid foundation—the continents' "basement" rocks.

In northeast North America, with Hudson Bay at the center, a huge flat plate of basement rock was formed, and it remains little touched by subsequent rock-forming events. Except for a lowland around the bay itself, the plate extends in all directions: north and east to Greenland, west into Manitoba, and south, roughly, to Lake Superior. This is the Canadian Shield, one of the great Precambrian shields of the earth's continents, more than a quarter-million square miles in area.

In gross morphology, the Canadian Shield is flat like a great dish, low at its Hudson Bay hub into which most of its surface waters drain. Over the immense length of time following hardening of the earth's crust to form the shield, surface erosion has leveled it to nearly a flat plain—the geologist's *peneplain*. Later, uplift and erosion by swift streams created locally rugged and fractured landscapes—rough terrain that much later was to be accentuated by glacial erosion.

Today, by foot, paddle, or portage, it's tough going. Still, stand on top of one of the local "mountains," and the horizon will appear flat—the line of the old peneplain.

The North Shore is not exemplary of the Canadian Shield. The great Superior Basin had not yet been created when the character of the shield was molded and altered. But, as we will see, the North Shore area was influenced greatly by the presence of these early Precambrian rocks.


The Precambrian Era has been roughly divided into several subunits, dated with radiometric techniques. The first of these—Archean, or Early Precambrian—included the development of the earliest rocks, both volcanic and sedimentary, which were later altered by heat and pressure.

The Archean closed with a great upheaval—the Algoman

*Granite cliffs, eastern shore*

orogeny, or Kenoran revolution—during which time high granite-cored mountains were formed by uplift and magma injected from the earth's interior. Only the granite roots of these Algoman mountains (named after the District of Algoma, Ontario) are left to us today. Even these roots present ranges of great hills, with their core of rock that did not originally reach the surface as magma but cooled underground. Some of the most prominent are the Giants Range in Minnesota and the sweeping Algoma hills along Superior's eastern shore. Their age: 2.7 billion years.

Later divisions of Precambrian time—the Proterozoic and its subunits—included different geologic events. In the early Proterozoic, great seas invaded the region, and the rocks that formed beneath the water had as their source the sediments eroded from highlands of yet older, Archean rock. This was the time of the Animikie sea, located in what is now the western and southern Lake Superior area. Sedimentary rock formed under the Animikie sea portended great economic significance, for these rock strata included iron-bearing formations of taconite, with 30 percent iron. Lapping against Minnesota's Giants Range, the sea left deposits that much later, only 100 million years ago, became concentrated by leaching and precipitation into the even richer ores—60 percent iron and more—of the Mesabi Range, deposits that once constituted the richest major source of iron in the world. The iron age: about two billion years.

Eventually the Animikie sea receded. Sedimentary rock that remained was uplifted and then eroded to a peneplain. Elsewhere on the continent, including the south shore region, there occurred much folding of crust and upward intrusion of magma. But in the area that was to become the Superior North Shore, the Proterozoic progressed through a period of relative quiescence continuing for nearly a billion years. The calm was to be suddenly broken by a period of volcanism and great crustal movement.

The time was the Keweenawan age, a term equated essentially with the time of the Midcontinent Rift System. In a broad arc, from an area in present-day Kansas, northeastward through Minnesota and Ontario, the rift curved through the region of present Lake Superior and continued southeastward into lower Michigan. Possibly it reached as far south as Kentucky. It was to leave mineral riches and natural beauty in great diversity along a

14

complex, 1,000-mile-long coastline of the earth's largest area of fresh water—Lake Superior.

The Keweenawan had begun quietly enough. In an early phase, predating the Midcontinent Rift, shallow seas and possibly large inland lakes had covered much of the Superior region and left sedimentary rock, traces of which remain mostly in the Lakehead and Nipigon regions of Ontario. Most prominent among these formations is the 1,400-foot-thick Sibley Group of red sandstone, shale, and limestone, exposed on Sibley Peninsula in Thunder Bay (from which the rock formation derives its name). In these sediments are preserved fossils of very early life—stromatolites, remains of primitive algae. The Sibley sediments also filled the 100-mile-long Nipigon, or Sibley, Basin. The Sibley's age: about 1.3 billion years.

But this early quiet period ended, and about 100 million years later began the most active period of volcanism—and the last—of North Shore geologic history. The earth's crust split apart in a gigantic crescent-shaped fracture, the Midcontinent Rift System. Lava burst from fissures in the widening crack until hardened layers accumulated to a great thickness, possibly 40,000 feet in the western part of the region, lesser in the east. The total volume of lava was about 100,000 cubic miles, more than thirty times the volume of today's Lake Superior itself.

Eruptions were repeated many times, located in different areas (perhaps eight major sites), so that the piles of lava layers resembled tilted, overlapping stacks of pancakes. Individual lava flows ranged in thickness from one foot to several hundred feet.

Over the area from which these immense masses of molten material had moved upward, the now incredibly heavy crust subsequently subsided, and the primordial Superior Basin was formed. The last and largest of the lava piles was situated between Isle Royale and Michigan's Keweenaw Peninsula; it later subsided to form one of Lake Superior's deepest basins.

The Keweenawan volcanism left a wonderful legacy along the North Shore, for the eroded lavas were to provide the foundation for a literal mélange of waterfalls and cascades in rushing rivers, vertical canyon walls, sheer coastal cliffs, and jumbled rocky strands.

15

*Location of the Midcontinent Rift System (shaded area), which
formed 1.1 billion years ago, creating at its northern crescent the
Superior Basin. (Adapted from Klasner, Cannon, and Van Schmus
1982, with permission of the authors.)*

The volcanic activity was associated with additional rifts and
faults, too. For example, a major crack in the earth's crust had
earlier extended northward, which, after later glacial erosion, resulted in present Lake Nipigon and the broad Nipigon Valley
(Chapter 15). Isle Royale, consisting largely of tilted edges of lava
flows, was uplifted along a fault between the island and the mainland (Chapter 16). Smaller faults occurred that were later transformed by erosion to beautiful stream valleys, like that of Ontario's Little Pic River.

Intrusions of molten rock also penetrated near the surface,
following fractures, small faults, and planes between strata of

older rock. These were diabase dikes (vertical curtains) and sills (horizontal layers) extending outward from dikes. Sills remain today with a thickness of several hundred feet. Together these formations of diabase account for perhaps the greatest diversity of landforms and the most striking of North Shore scenery, for the hardness of the dikes and sills differed from the older rock into which they intruded.

Features resulting from diabase formations include waterfalls like High Falls on the Pigeon River and the upper Chippewa Falls on Harmony River; resistant wave-sculptured cliffs like Silver Cliff in Minnesota and the Palisades in the Nipigon Valley; and scenic points that extend into Lake Superior, like Minnesota's Pigeon Point and Hat Point. The formation of the Sleeping Giant, off the city of Thunder Bay, has on its near (west) side the highest sheer cliff on the North Shore—800 feet. Dikes resulted in deep gorges and canyons like those of the lower Montreal River, where the dike was *less* resistant than the surrounding rock. Spectacular Ouimet Canyon, east of Thunder Bay, was eroded by glacial meltwater along fractures in a diabase sill.

Diabase sills also left towering mesas and islands like 1,000-foot-high Mount McKay and Pie Island at Thunder Bay and the intriguing formation of the "Sea Lion" projecting from Sibley Peninsula. Most prominent of all, perhaps, are the towering ridges and cliffs, alternating with deep, flat valleys, in the Grand Portage-Thunder Bay-Nipigon Bay region, the "Nor'Westers Mountain Range."

A curious anomaly exists in northeastern Lake Superior, unrelated to the Keweenawan volcanism and the Midcontinent Rift System: the Slate Islands. Located off the shore near Terrace Bay, they are thought to have resulted from the impact of a meteorite, occurring sometime after Keweenawan events.

Collectively, the lava flows from the Keweenawan eruptions are known along the Minnesota coast as the North Shore Volcanic Group. Lava remnants occur along the Ontario shore, too, in similar profusion: for example, on the north, the lavas on Black Bay Peninsula, and to the east, the lavas of Gros Cap, Mamainse Point, and Michipicoten Island. Along the eastern shore, however, much of the Keweenawan lavas lie beneath the waters of Lake Superior. Their age: 1.1 billion years.

When the Keweenawan volcanism ended, major earth-shaping events on the North Shore ceased. Another half-billion years were to pass before the Precambrian closed, but there was no more major volcanic activity. In late Keweenawan times, great freshwater lakes formed, fed by braided streams flowing toward the axis of the rift, but these fresh waters left little evidence on the North Shore. Sedimentary rock, such as the red and yellow Fond du Lac and Hinckley sandstones, were formed in these late Precambrian lakes or by braided river systems; these rocks now barely touch the North Shore at the western end of the lake. One part of Isle Royale contains a remnant of sandstone and conglomerate. Other sedimentary rock from these times, such as the later Jacobsville Sandstone, underlying much of the Lake Superior Basin, appears at the extreme eastern end, and these rocks are the youngest on the North Shore. The red Fond du Lac, Hinckley, and Jacobsville formations are thought to be responsible for the red color of clay in western Lake Superior sediments, deposited when later glaciers ground the sandstone to a fine flour and copious meltwaters were turbid with red clay.

And so the Precambrian closed quietly, and the North Shore was left to mature under the relatively milder influences of wind and water for almost the rest of the earth's history. The time: a half-billion years ago.

The name *Precambrian* implies that a time called the *Cambrian* should follow—and it did. The end of the Precambrian signaled a major change to the next era—the Paleozoic—and the Cambrian Period is the first subdivision of the Paleozoic Era.

And what was this major change between eras? The answer is found in the translation of the Paleozoic name: *ancient animal life*. The major change, then, was a transition from initial rock formation with only nascent life to the rapid evolution of increasingly modern organisms. In the warm, placental seas of the Cambrian Period, a half-billion years ago, animal forms evolved in ever greater complexity and flourished in fantastic profusion. The Cambrian seas and the fossils they left, however, will not concern us on the North Shore, for no sediments from Paleozoic time remain on Superior's northern coast.

One more major geologic event was to leave a dramatic imprint along the North Shore: the glaciation of the Pleistocene Epoch, the great Ice Age. Compared with the vast stretch of time that had gone before, this event occurred much more recently. And while glaciers did not disrupt the earth's crust to the same extent as did volcanism, glaciation was responsible for a final melding. Because the last glaciation occurred so recently in our past—10,000 years ago—its effects today are in many respects even more visible than those of the earth's fiery genesis.

During the early part of the half-billion years following the Precambrian, much material was probably eroded from the highlands surrounding the Superior Basin and washed down into it. Upon the red sandstones laid down during the late Precambrian then came immense volumes of rock debris to form a wide, flat plain. A large river perhaps flowed across the plain, eastward toward what was to be the Atlantic Ocean, maintaining a broad valley.

Highly significant to the Superior region later on, this valley constituted a great lowland that was to guide a tongue of glacial ice, through the valley and on southwestward far across the state of Minnesota, largely responsible for present land forms in the central part of the state.

Glaciation of the Pleistocene Epoch began about 2.3 million years ago when the earth's climate turned colder. At several centers in northern North America, the summers were cool enough so that not all of the previous winter's snow melted. And so the snow accumulated, year after year, century after century, until it became hundreds of feet thick. The tremendous weight pressed the lower accumulations of snow to ice, then the ice to a plastic state; the ice began to move away from these centers of accumulation, southward across the North American continent. The weight of ice, eventually up to two miles thick, was to exert tremendous pressures upon the earth's crust, depressing the land many hundreds of feet.

Glaciation of the Pleistocene occurred in at least four distinct stages, with warm, interglacial intervals between, during which the climate was much like that of today, perhaps a bit warmer. The four stages—Nebraskan, Kansan, Illinoian, and Wisconsinan—were named after the states in which terminal remains of glacial debris are now most evident and where the principal respective geological studies were made. The first three stages will not concern us on the North Shore, for their remains have been destroyed by the last. Rather, it was the Wisconsinan Stage, beginning about 100,000 years ago, and ending only about 10,000 years ago, which had the greatest visible effect on Lake Superior and the North Shore.

The movement of glacial ice was not a regular, linear progression across the continent, but rather, as previously suggested, in distinct tongues or lobes. Lobes advanced and melted back, sometimes from different centers and thus in different directions. While one lobe melted back to expose an area of land, another progressed from a different direction toward the same area, carrying different materials and leaving different kinds of rock debris (thus allowing us to determine what happened). The movement of ice in lobes occurred because each followed a preexisting lowland, through which the ice was guided by higher ground on both sides. This was particularly true in later stages, when ice was thinner and surface topography had an increasing influence in guiding glacier movements.

The Superior Basin constituted one of these major lowlands. When ice moved southward over the continent, some formed into the Superior Lobe, guided by the highlands of hard Keweenawan lavas on north and south sides. The lobe progressed southwestward, continuously fed by faraway northern snowstorms.

The Superior Lobe advanced and melted back four times during the Wisconsinan Stage, each time leaving evidence of its presence in the form of distinctive rock debris, or glacial drift. Sometimes the Superior Lobe competed with other lobes from the north and west, sometimes coalescing with them, sometimes mixing and interchanging their unique sediments as one lobe advanced over the drift left behind by another. The Superior Lobe probably did not melt all the way back out of the Superior lowland during the several retreats, but each time it advanced it continued to quarry and abrade debris from rocks formed a billion years ago in the Superior Basin—and leave that rock debris farther on as glacial drift.

A major effect of all this excavation of Precambrian material out of the Superior lowland was deepening of the basin. Exactly how deep we do not know; one experimental drilling into Lake Superior's bottom went down to more than 1,000 feet below sea level without hitting bedrock. With surrounding land elevations today more than 1,000 feet above sea level, it appears that at the time glaciation ended the rock basin would have been at least 2,000 feet deep.

When the terminus of an ice lobe melted at about the same rate as the ice moved ahead, the forward edge remained in about the same location. But this meant that much rock debris carried by the ice was deposited at one place, as northern ice continued to arrive and melt.

Thus, the glacier lobe acted like a giant conveyor belt carrying boulders, gravel, sand, and silt through the lowland to dump it all in great heaps at the glacier's margin. The result was a range of great hills—terminal moraines. It was the Superior Lobe that left a complex of moraines in an impressive landscape—the lake and hill country—in central Minnesota.

At one stage of the Superior Lobe's advance, it pushed out of its basin not only forward toward the southwest, but also up and outward laterally, to the west and northwest, and formed a lateral moraine high up on the northern edge. This is the North Shore Highland Moraine, deposited upon older drift and the Keweenawan lavas that form the North Shore, from Duluth northeastward for more than 100 miles. The Highland Moraine now overlooks Lake Superior from heights up to 1,500 feet.

But after forming a terminal moraine, when the rate of melting exceeded the rate of forward movement, the edge of the ice receded—and meltwater, impounded by the moraine in front and the existing ice in back, formed a great pool: a proglacial lake.

The Superior Lobe formed proglacial lakes a number of times as it retreated. In sporadic readvances, ice eroded previously settled lake clays and deposited them again farther southwestward in more moraines.

Upon final retreat of the ice, the lakes it formed left evidence which remains today—in lake sediments and their flat topography, the courses of now-dry outlet streams, and prominent gravel beaches, terraces, and old strandlines. The Superior Lobe left a complex historical record, difficult in its unraveling, but fascinating in the variety of its detail.

Lake Superior, and indeed all the Great Lakes, were once larger glacial lakes, until outlets were uncovered that eventually allowed drainage through the St. Lawrence River to the Atlantic Ocean.

        The first of the proglacial lakes was relatively small, just at the tip of the Superior Lobe west of the basin, about 12,000 years ago. A progenitor of Glacial Lake Duluth, about 450 feet higher than that of present Lake Superior, it drained southwestward in now abandoned outlets toward the Mississippi River.

The Superior Lobe continued to melt eastward. It became thinner, edges retreated toward the lake's center, and more and more land along its sides was uncovered. The tip of the lobe retreated back up the basin.

Then another outlet channel was uncovered by the melting ice in what is now northwestern Wisconsin. This channel led south and west, also to the Mississippi, and it was in this channel that Wisconsin's Bois Brule River and the St. Croix River later developed, the waters in the Brule still later reversing to become tributary to Lake Superior. With more ice melted to water, the larger Glacial Lake Duluth was established, its level still much higher than that of present Lake Superior. The waters of Glacial Lake Duluth poured out in immense volume, carving the spectacular St. Croix Valley.

The climate continued to warm, and other lobes occupying the Lake Michigan and Lake Huron basins began melting back, too, forming other large glacial lakes.

Several times ice retreated and readvanced. New outlets were uncovered to the east and south but were covered again with ice as glaciers moved forward. Between 10,000 and 9,000 years ago, the Superior Basin became finally cleared of ice, and the land along the northwest edge of the basin lay open. The final wastage of the Superior Lobe from the Superior drainage cleared the basin, creating Glacial Lake Minong, which was slightly larger than present Lake Superior.

Now between Glacial Lake Minong and eastern lobes of Glacial Lake Agassiz to the northwest lay an open corridor of newly uncovered land. And it was probably at this time—between 9,500 and 9,000 years ago—that plants and animals first became permanently established on part of the North Shore.

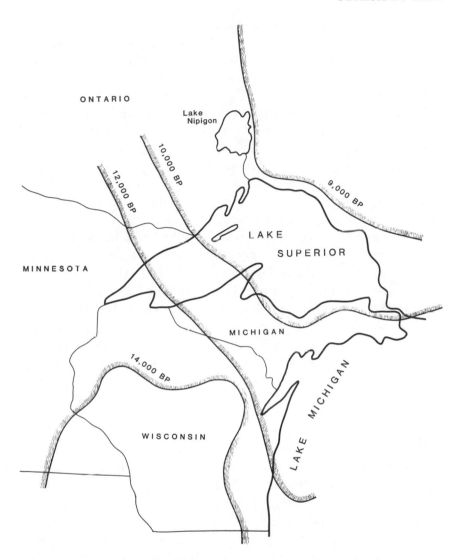

*Retreat of glaciation northeastward across the Superior Basin.*
*Present Lake Superior outline shown. BP = years before present.*

The warmer climate continued, and soon ice retreated northward beyond the Superior-Hudson Bay divide.

With ice lobes gone from all the Great Lakes basins, eventually waters in the Superior, Michigan, and Huron basins coalesced. A vast interconnected glacial lake formed, which existed for a very long time and at many different levels as different outlets were uncovered, uplifted, or eroded. Its ultimate drainage was to the Atlantic, not the Mississippi, and the development of its final release to the sea, with establishment of the present Great Lakes and their drainage out the St. Lawrence River, only about 2,000 years ago, is a complex and fascinating tale, not told here.

But changing water levels and outlets affected the North Shore of Lake Superior in many important ways. Succeeding water levels left successively lower strandlines and beaches along the shore, visible today in terraces and gravel beach ridges. The red clay and silt that were derived from the glacial crushing of Precambrian sandstones, later deposited under glacial lakes, remain now on much of Lake Superior's coastline and in North Shore river gorges. Some of the more prominent strandlines include glacial lake beaches near the city of Duluth (from which the glacial lake derived its name); the flat delta of an earlier Kaministiquia River, at Thunder Bay, deposited when the river flowed into glacial lakes at higher elevations; and other beaches and terraces at Terrace Bay, Batchawana Bay, Michipicoten, and Sault Ste. Marie. The beaches at Terrace Bay are particularly noticeable, accounting for the name of the town.

Upward tilting of land in the northeast, due to relief from the weight of the ice, led to further changes in lake levels. Tilting is still occurring. And although subsequent changes in lake levels and old strandlines seem very small to us now and far removed from our ancestral Lake Superior, they are reminders of just how close modern civilization is to the events of the great Ice Age.

When glacial ice finally left the Lake Superior region about 9,000 years ago, the climate had begun to warm more rapidly. Many large masses of stagnant ice remained covered by glacial drift, which insulated and delayed their melting. Later, when these ice blocks did melt, the overlying material collapsed to form lake depressions. Such ice-block lakes account for most of the lakes in the Great Lakes region—but not on the North Shore. Here glacial drift was thin, and most inland lakes are the result of glacial scouring of older river valleys in Precambrian rock.

Tundra and forest followed the ice northward, and successional changes are recorded now in fossilized remains in the sediment of lakes and bogs. Pollen grains in different layers of these sediments offer us a chronological record of changes in vegetation over this 10,000-year period.

Today, far to the north, remnants of glacial ice persist, like those on Canada's Baffin Island. Will they readvance again, leading to another ice age?

Theories of glacial cycles predict that, indeed, they will.

And so springtime, following the long winter of the Pleistocene, came at last to the North Shore. Glaciers had held the Superior Basin in their icy grip, on and off, for more than two million years. Now a moderate climate with abundant fresh waters, sheltering forests, and diverse populations of near-modern mammals, birds, and fishes set the stage for one of the most profound ecological events of northern North America and the Superior country: human colonization.

# T W O

# *The Early Americans*

S hadowed in prehistoric millennia, the exact time of arrival of humans on the North American continent remains concealed. The date is today a point of much controversy. Some estimates place it as long ago as 100,000 years, although that is largely conjectural. The major migration probably occurred about 20,000 years ago.

At that time, the last glacial stage of the Pleistocene Epoch, the Wisconsinan, was near its peak. So much of the earth's water had been taken up in ice that the ocean surface was more than 300 feet lower than its present level, exposing in the northern Pacific a broad land connecting Asia and North America—a land we now call Beringia, the Bering Land Bridge, nearly 1,000 miles wide.

Across this corridor wandered many large Asian mammals: the giant bison, muskox, camel, woolly mammoth. And following these game animals came *Homo sapiens*—modern humans, like ourselves—to spread eastward in small hunting bands. They could not have known they were about to colonize a new continent.

We know that by 11,000 years ago humans occupied much of southern North America. Some, those who will attract our attention because of our interest in the North Shore of Lake Superior, lived on the High Plains east of the Rockies. They hunted the large mammals that grazed the prairies, and they spread eastward and northward as the glaciers retreated.

By about 10,000 years ago, considered the end of the Pleistocene, glaciation in North America approached its close. Much of the world's ice had melted, and ocean levels rose again. The Bering corridor was inundated by the rising sea, and so this route of human dispersal between the two continents was sealed shut. Henceforth, any human intercourse between the two hemispheres would necessitate travel by water across broad seas—admittedly, an event that our study of prehistory has not ruled out.

The Asians that migrated across Beringia would spread across the continent, to the Great Lakes and the Superior North Shore, and beyond, to the shores of the Atlantic Ocean. They would develop their own culture as native Americans, distinct until millennia later when adventuring Europeans would visit North America in their sailing ships.

In the Great Lakes region, because of fluctuating levels in glacial lakes, the migrations and settlements of human populations left a tangled record. With the Ice Age drawing to a close, human cultures developed in accordance with glacial retreats and minor readvances, with the formation of vast proglacial lakes, and with the final uncovering of land. Pressing against the ice perimeter, living along shores of glacial meltwaters, invading and occupying newly exposed lakebed plains, these Ice-Age Americans hunted and spread east and north. Their colonization of the Great Lakes region depended most directly on animal response to vegetative succession, which in turn depended on the changing climate.

The first vegetation on land newly uncovered by retreat of the glaciers was similar to present-day northern tundra. An open landscape, too cold for trees, the tundra comprised only dwarf birch and willow, and arctic grasses. Characteristic of a cold northern climate, tundra followed in the wake of the melting ice.

The periglacial wildlife of the Pleistocene included large mammals that could withstand the cold—a megafauna. Some species became extinct as climates changed and human hunters exacted their toll. But three large mammals of the tundra re-

*Ojibway canoe*

mained, at least for a while: the woolly mammoth, muskox, and the barren-ground caribou, although the last made annual migrations from tundra to southern forested lands in winter. These were grazing animals that depended upon open grasslands and the low vegetation of the treeless tundra, or at least upon open boreal woodlands.

At the western end of the North Shore, tundra disappeared about 11,000 to 10,000 years ago, probably a bit later farther east. In its final few hundred years, the tundra changed to a shrubland of dwarf birch.

After about 9,500 years ago, when the western North Shore became free of ice for the last time, came the first of the coniferous trees: black spruce—vanguard of the boreal forest. Boreal forest was to dominate the vast lands north of Superior up to modern times.

The mammoth became extinct; the muskox moved north;

but the caribou remained as the principal large mammal in the black spruce.

The climate gradually warmed, and about 9,000 years ago from the south came trees of successively warmer climates. Deciduous trees such as white birch and alder spread northward as the climate changed and glacial ice retreated into the far north.

About 8,000 years ago the climate warmed dramatically, and some conifers were replaced by other species, especially the deciduous oak and elm. The white pine arrived about 7,000 years ago from the east; it spread along the southern shores of Lake Superior and up around the western end to the North Shore. The climate on the North Shore became even warmer and drier than it is today.

This mid-postglacial warm interval, lasting 4,000 years, was responsible for many vegetative successional changes. During the latter part of this period, oak forest and even prairie invaded some of the western parts of the North Shore, albeit temporarily, and black spruce retreated into the far north.

About 5,000 years ago, the climate reversed again toward the more humid, cooler state of the present. Oak and prairie retreated westward, and white pines followed; the spruce forest returned to the North Shore.

With changes in vegetation came changes in wildlife populations as well. Some species became extinct, although the woodland caribou remained in the boreal forest. Moose had not yet arrived; the white-tailed deer prospered in the mixed conifer-deciduous forests to the south. All were to be affected greatly later by anthropogenic changes in their environment. Coldwater fishes, such as lake trout and whitefish, colonized the deep, clear waters of newly formed lakes and rivers in the glacier-scoured Precambrian rock. Northern pike and walleyes, of somewhat warmer waters, followed later.

For the last 4,000 years, until the time of white settlement and logging, woodlands on the North Shore have remained essentially the same. At both the western and eastern ends persist mixtures of both conifers and deciduous trees. But about in the center of the shore, from a little west of Thunder Bay eastward and south around the lake to near Wawa, the boreal forest extends down from the north, wraps around Lake Nipigon, and lies tight against Lake Superior. Here still are dense concentrations of black spruce, whose ancestors once spread northward behind glacial ice and tundra.

As suitable environments thus expanded northward, so then did prehistoric peoples, basically remaining as hunters, filling niches at the limits of their natural world—as that world changed from ice to water, from open tundra to shielding forest.

The prehistory of human occupation in the Great Lakes region has been divided into four major periods, each with some distinctive attribute of human culture. These time periods are artificial, however, to the extent that they are the archaeologists' inventions—albeit based on real differences in human culture between periods—derived from artifacts found in diggings at occupation sites.

The prehistoric periods and their associated cultures on the North Shore are as follows: (1) Palaeo-Indian period with the Plano culture, 11,000 to 7,000 years ago; (2) Archaic period and the Shield Archaic culture, 7,000 to 3,000 years ago; (3) Initial Woodland period and the Laurel culture, 3,000 to 1,000 years ago; and (4) the Terminal Woodland period and its Algonkian culture, 1,000 years ago to the early 1600s. This last period brings us up to the time of European contact and historic times, when written records were kept—and when the cultural development of early North American peoples changed drastically.

Earliest archaeological evidence of humans in North America is assigned to the Palaeo-Indian period, beginning about 11,000 years ago. The Palaeo-Indians—a term meaning ancient Americans—descending from the migrants who crossed the Bering Land Bridge, spread quickly eastward across the continent. But to the north they met the barrier of the glaciers, and in the Great Lakes region, where melting ice was at that time forming vast proglacial lakes, their eastward advance was temporarily blocked.

But as retreating glaciers permitted, some peoples advanced toward the North Shore of incipient Lake Superior. About 10,000 to 9,000 years ago, when ice retreated from the Superior Basin and Glacial Lake Minong existed, the people of the Plano culture

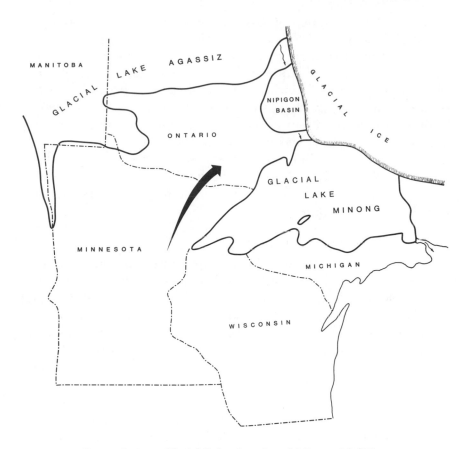

*Passage between Glacial Lakes Agassiz and Minong 10,000 to
9,000 years ago. Uncovered for the last time, this passage permitted
colonization of the western North Shore area successively by tundra,
game animals, and humans (bold arrow). Small arrows indicate
spillways of glacial meltwater.*

spread northeastward, through an open passage between Glacial
Lake Agassiz on the west and Glacial Lake Minong on the east, to
occupy part of the northwestern coast of the Superior Basin—the
North Shore's first human occupation. Palaeo-Indian artifacts of
taconite—scrapers, blades, hammers—attest to occupation in the
Lakehead area between 9,000 and 8,000 years ago.

   Although the Palaeo-Indians lived along the water's edge at
the time, their evidence was left at higher elevations and farther
inland than present lake margins, because glacial lakes of the time

were larger and higher—Glacial Lake Minong being about 150 feet higher than modern Lake Superior.

The Palaeo-Indian period continued for about 4,000 years, and during this time, cultural changes took place in accordance with the succession of tundra and boreal forests north of Lake Superior. Using spears with stone dart heads, the Plano people hunted caribou on the tundra and in the open boreal forest, later in woodlands. Fishing was probably not at first an important element in these early Americans' food economy; cold and nutrient-poor waters, turbid conditions of glacial meltwaters, and unstable conditions would have militated against high fish productivity, which was to come later.

Changes in human culture followed climatic changes into the next period—the Archaic—7,000 to 3,000 years ago. In the north, the people of the Shield Archaic culture (that is, on the Canadian Shield) adapted to the cool forests, mainly by hunting the woodland caribou as their staple food. But they developed new weapons and tools for hunting and processing other and smaller game: bear, beaver, snowshoe hare, waterfowl. When fish populations developed, the Shield Archaic people fashioned fishhooks and nets for fishing. They probably traveled by birchbark canoe in summer, and by some form of snowshoe in winter. As far as we know, Shield Archaic people were the first to have dogs—the first domestic animals—in the Great Lakes region. There is abundant evidence of widespread trade with cultures elsewhere on the continent. More advanced tools were developed, such as the ax, adz, and gouge.

It was within this Archaic period also that the Old Copper Culture appeared, extending into part of the next period, the Woodland. Mined from copious deposits on Isle Royale and its southern counterpart, the Keweenaw Peninsula, and a few other places around the North Shore such as Mamainse Point, this malleable metal was shaped into weapons, tools, and fishhooks and gaffs. Pure veins were pried from Precambrian lavas, heated, and hammered into the tools of their culture—the first experiments in metallurgy in the New World. Although copper use extended broadly across the North Shore, it did not extend much farther northward; northern Shield Archaic populations were separated by great distances of wilderness from Lake Superior and obtained copper only in occasional trade.

The Shield Archaic culture coincided, roughly, with the warm mid-postglacial interval. The milder climate might have contributed to the people's cultural development, for they could devote more time and effort to tools, living conditions, and travel, rather than wholly to survival. The more amenable environment probably resulted in greater populations and expansion northward.

One more important introduction to the early Americans' culture—the bow and arrow—was truly revolutionary, enabling them to hunt wild game more efficiently. The bow and arrow appeared in the late Archaic period, about 3,600 years ago.

Following the Archaic period, a final prehistoric period—the Woodland—is recognized by archaelogists. In the northern Great Lakes region, the Woodland was further divided into Initial and Terminal periods. The Initial Woodland extended from 3,000 to 1,000 years ago, the Terminal Woodland from 1,000 years ago to the arrival of white Europeans, who drastically changed the early Americans' life.

While the transition from the Archaic period to the Initial Woodland period was gradual, one important change was preeminent: the introduction of pottery. The early culture of the Initial Woodland was largely the same as the Archaic except for this discrete change. And although the idea of pottery was introduced from southern and eastern peoples, the pottery of the Initial Woodland, north of Lake Superior, was developed into distinct styles.

The people and their life-styles in the vast, northern coniferous forests are termed the Laurel culture in this Initial Woodland period. There was more far-ranging travel for trade in distant regions, exchanging North Shore copper for obsidian from Wyoming and seashells from the Atlantic coast. And in southern and western parts of the Great Lakes region, including the western part of the North Shore, wild rice became an important dietary element in the Laurel culture, beginning about 1,200 years ago.

The Terminal Woodland period, beginning about 1,000

years ago, included in the region around Lake Superior the Algonkian culture, approaching historic Indian life-styles. The term *Algonkian* refers to a language connection that developed in three major regions, eventually resulting in the three major historic clusters of Indian bands. These bands were represented by the Cree, to the west and far north of Lake Superior; the Algonkins, east of Lake Superior and Georgian Bay; and the Ojibway (or Chippewa) around the eastern end of Lake Superior.

The three Algonkian cultures were not discretely separated, and there was considerable travel and communication among the three. We know this primarily from the pottery remains. Since family descent was patrilineal, and marriage was prohibited within a family group, wives were often drawn from other bands and areas. Women made the pottery, and men manufactured weapons and tools. Thus, at the site of a local band, remaining artifacts included a variety of pottery types but a consistency of tools and weapons. This interchange also resulted in the broad similarity of the Algonkian language. Still, the three major sub-cultures differed sufficiently to be distinguishable on the basis of archaeological remains.

In the eastern North Shore region, the Ojibway ("people whose moccasin seams are puckered") lived a life-style of hunting, fishing, and harvesting wild fruits and seeds. The gill net for fishing, a major technological advance in harvesting the bounty of Lake Superior, appears to have been first used during this time. Together with seines, invented earlier, the gill net provided the Algonkian culture with a fishing technology for which it became widely noted.

Hunter-gatherer cultures were not normally warlike; their semi-nomadic life led neither to the accumulation of property nor to the coveting of their neighbors'. Rather, it was farther south and east, where agriculture was developed and permanent villages established, that stored foods and manufactured goods invited armed forays.

In modern terms, population density was sparse north of Lake Superior, with humans living in balance with their harsh environment and infertile resources, in essentially a predator-prey relationship.

The study of prehistory is in the scientific realm of the archaeologist, who pieces together the puzzles of remaining artifacts and other data. But in the acid soils of the North Shore, fragments of only rock and metal remain; organic materials of bones, leather, and wood have long since disappeared. The archaeologist's job is thus made more difficult in these northern regions than in more southern climates and soils.

But here and there, at old habitation sites, burial mounds, and campgrounds—especially where the ashes of repeated camp and cooking fires have alkalinized the decaying material—some evidence remains. Layer upon layer, culture upon culture, the evidence discloses the development of the lives and life-styles of these early Americans along the North Shore. In lower layers still rests the evidence of early cultures, and successively higher strata show signs of steady cultural change. In the uppermost layer appear artifacts of European origin: trade goods of utensils, tools, ornaments, weapons.

Of human dispersal around the world—from the eastern hemisphere across the Bering Land Bridge and the Pacific Ocean 20,000 years ago, and from Europe westward across the Atlantic Ocean by sail only a half-thousand years ago—the cycle was completed. These last cultures of early Americans along the North Shore were about to change dramatically.

In the beginning of the historic period on the North Shore—the early 1600s—white Europeans found first the Ojibway living essentially around the eastern end of the lake and along the North Channel of Lake Huron. Some were in part of the Upper Peninsula of Michigan, south of Lake Superior, but most lived along the North Shore up to a little beyond the Michipicoten area.

The rapids, or sault, emitting from the big lake was the focus of the Ojibway's culture, for it was here that they had developed a highly productive fishery for lake whitefish, a major resource for the Indians' sustenance. The French name for the tribe was thus *saulteurs*. In the Lakehead area, around Lake Nipigon, and up through the vast lands to the north and the Hudson Bay area, were the Cree; the Assiniboin occupied the northwestern shore of Lake Superior and its hinterlands; and the Dakota, or Sioux,

*Distribution of Ojibway (shaded area) and other tribes at time of contact with Europeans, in the early 1600s. As the fur trade progressed westward during the next two centuries, the Ojibway dispersed around both north and south sides of Lake Superior (arrows), displacing the other tribes. (Redrawn from Bishop 1974, with permission of the author.)*

lived around the western tip of the lake, as far east as the Pic River on the North Shore.

As white explorers and fur traders gradually moved westward along the shore, the Ojibway moved westward, too. Eventually, the Cree were pushed into northern woodlands, and the Assiniboin westward toward the prairies. Ojibway south of Lake Superior also pushed westward, ultimately to replace the Dakota in that area.

This westward movement of the Ojibway appears to be a direct result of the fur trade, and of the dependence of the natives upon the trade, as explorers, traders, and white civilization moved westward. By the middle of the eighteenth century, the Ojibway occupied the entire shore, both north and south, around Lake Superior.

The fur trade in its heyday of the late 1700s came into contact primarily with the Ojibway. Records of Indian culture kept by white traders, scanty as they were, thus were much more complete for the Ojibway than for the other tribes who had occupied parts of the North Shore in prehistoric times.

The Ojibway's way of life on the North Shore was mainly a hunting-based culture, but three different subsistence strategies can be identified at various locations along this shore.

First, along the northeast side of Lake Superior fishing was most important to the Ojibway. They fished mainly for whitefish and lake trout in autumn, using gill nets; if it was late enough in the season, fish were frozen for later use, or dried and smoked for preservation. Fishing on this part of the Superior coastline was facilitated by the much greater occurrence of protected bays, shoals, and beaches, in contrast to the more rugged and hostile character of the northwest side of the lake, with its straighter, rocky coastline and deep offshore waters. Netting whitefish in the rapids of the St. Marys River was a major part of the fishing activity near this part of the shore. Great Lakes aboriginal fishermen are considered as skillful in their fishing activities as were any on the North American continent, especially in the manufacture and use of gill nets. But hunting, too, was a vital part of economic strategy in this area, as was maple sugaring in southern parts of the shore.

Second, natives living along coastal edges in the Lakehead region and much of the northwest coast depended almost solely on hunting. The dense conifer woodland of the boreal forest offered little in the way of deciduous fruits and seeds. The woodland caribou was the staple large game, the principal species large enough to make hunting energy-efficient, but smaller game animals were taken as well. In times of winter scarcity, the Indian depended heavily upon snowshoe hares. In winter, meat could be frozen for storage, but there was little other means for preservation, and consequently spring was often a time of food scarcity and hardship, before seasonal foods of summer became available.

Third, along the far western part of the coast occurred a portion of a wild rice culture, although most of the wild rice region lay to the west and south of Lake Superior. Use of wild rice began here about 1,200 years ago, possibly earlier. Wild rice was stored, when parched and treated, for later use in lean times, a development that enabled the human population of this area to increase. But, here too, hunting was also an essential element in human food strategy.

The total native population around Lake Superior was something like 20,000 to 25,000, about one person per two square miles on the average, although densities were undoubtedly lower on the North Shore than in areas farther south. Below Lake Superior, some agriculture was practiced, and primitive though it was, cultivation of richer soils and growing of some of their own foods permitted more permanent settlements and higher densities. But in the north, with its infertile Precambrian rock base and thin glacial soils, with short seasons and low temperatures, agriculture was virtually impossible. Food acquisition strategies never progressed much beyond the hunting and fishing stage.

In a society based primarily on hunting, the Ojibway's life was nomadic. They were organized into fairly independent bands that occupied given geographic areas, each band including many family groups, or clans. A typical example might have been a band of 600, composed of thirty family groups or clans of twenty persons each, using an area of 1,200 square miles. At the time of contact with Europeans, the Ojibway population probably numbered about 4,500. Their housing consisted of dome- or cone-shaped wigwams with rounded tops, large pieces of birch bark covering bent-over willow saplings, insulated with moss.

In winter, from isolated family groups, the men traveled locally on snowshoes for hunting, using the bow-and-arrow, spears, snares, and clubs. With their skill in the use of gill nets, it is probable that the Ojibway fished their nets through the ice in winter, too.

Winter was a time of trial. Dependent upon wild game, the resident unit that could be so supported was small, perhaps one or two families, for the forest's wildlife was not sufficiently abundant to support more in a given area limited by difficulty of winter travel.

But in the spring, at least in the deciduous forest at the southeast end of the North Shore, the making of maple sugar was a welcome relief from winter's travails. Many families assembled for what amounted to a spring festival at the annual sugar bush. Maple syrup and sugar alone sustained much of the Ojibway population in this part of the shore for a critical part of the spring, before frozen coast and rivers could be traversed.

In the summer, travel was easier, mainly by birchbark canoe along the myriad watercourses of lakes and streams, opening up more far-ranging hunting and trade. The wigwam's bark coverings were rolled up and carried to the next camp. Family groups, released from winter's confinement, congregated on larger, traditional campgrounds, usually along some major waterway, for intensive fishing. Great numbers of whitefish were taken in the St. Marys rapids. In late summer and fall the Ojibway gathered the wild harvest of meadow, marsh, and forest plants, and in late fall the spring trek was traced in reverse back to winter family camps.

The arrival of the Europeans in the 1600s sent a cultural shock wave across the North Shore, east to west. The explorers, missionaries, and traders irrevocably changed the natives' lifeways, at first mainly in material goods. Arrow stone tips were replaced by brass, stone knives and axes by iron, traditional pottery by brass kettles—and later, primitive weapons by firearms. Trapping of fur-bearing mammals, primarily beaver, using the white man's efficient iron leg traps, eventually became the major occupation, trading pelts for manufactured utensils, fabrics, tools, and weapons.

The nomadic life, harsh but self-sufficient, ended, and a dependence upon the Europeans' and their trade goods began.

The Ojibway and Cree north of Lake Superior survived the effects of the Europeans' intrusion better than did tribes farther south. In the north, theirs was a land of wilderness—harsh climate, thin and infertile soils, short summer growing seasons—almost totally unsuitable for the more agrarian culture of the whites. The fur-trapping era was a rich one for the European traders, but it was short, and it left no permanent settlement.

In 1850, the Ojibway of Canada signed away their portion of the North Shore, except for three small reserves near Fort William, Lake Nipigon, and Michipicoten, in Canada's Robinson-Superior Treaty. And in the Minnesota portion, the United States Treaty of 1854 extinguished the natives' ownership of land, except for Grand Portage Reservation near the international boundary.

Yet the Ojibway survived along the North Shore, even more distinctly farther north. The rugged beauty of the North Shore survived, too, though not in the pristine wilderness condition of the prehistoric Ojibway.

For 10,000 years, from their initial occupations along great glacial lakes to occupations dependent on iron traps and gunpowder from Europe, the early Americans were the sole human occupants of the North Shore.

That era ended with the inclusion of European artifacts in the uppermost stratum of the archaeologist's diggings.

# THREE

# *The Northwest Adventurers*

When and how a North Shore Ojibway first viewed the European explorer, we can only conjecture. Along a shaded, woodland trail, perhaps, he was stopped in shock at the sight of a band of those strangely dressed intruders; perhaps, from some rocky outpost along the shore, he watched with a mixture of curiosity and anxiety a brigade of canoemen in their birchbark craft, much like his own but larger, as they dipped their crimson paddles in unison into the crystal waters of one of Lake Superior's rare littoral shoals.

More likely, the Indians had already heard from eastern tribes of the Europeans' approach and then traveled to the white men's camp to see for themselves who it was that came adventuring into their familiar world of forests, streams, and rocky coast.

Certainly the two races—separated in time for scores of thousands of years—now met with curiosity and, at least at first, amicably. The time was in the early seventeenth century, possibly about 1615. The leader of the first pathfinders: Étienne Brulé, one of Samuel de Champlain's agents, from New France on the lower St. Lawrence River.

Brulé was the vanguard of a vast army to follow into the northern Superior country. Its numbers would increase at first only very slowly; in 100 years it would expand in a crescendo of

*Middle Falls, Pigeon River*

activity, both noble and commercial; in 200 it would reach a climax; and, in another 50, it would decline, with only faint traces left of its forest trails.

The invaders included, first, audacious French adventurers seeking a short route to the Orient, by way of a mythical Northwest Passage and, of course, *la gloire* that would follow discovery.

Later, as these vanguards of the western forest grew in numbers, they were joined by representatives of Europe's courts intent upon claiming vast but unknown lands for their kings; black-robed priests intent upon saving the heathen's soul, sometimes at the expense of martyring their own; government emissaries who ventured among the warring early Americans to arbitrate and placate.

Above all, however, this Northwoods army consisted of fur traders and their laborers, the all-consuming object of whom was the beaver—*Castor canadensis*—builder of dams and canals throughout the North American forest, unfortunate enough to possess a body covering of fur so fine and so suitable for making fashionable hats as to be coveted by all the swains, soldiers, and gentlemen of civilized Europe.

The tale begins along the wild shores of a continent new to the Europeans, where a mighty river, eighty miles wide at a point where it could no longer be called a river but a gulf, eventually entered the stormy seas of the north Atlantic Ocean.

Toward this coast of North America in the early years of the sixteenth century, European powers cast covetous eyes and directed their sailing ships. The time was only a scant few decades after Columbus's epic voyages. Spanish ships and conquistadores probed and penetrated southern estuaries and Caribbean straits; French fishermen plied the cold waters of the north Atlantic's shoals and banks; conservative England planned for colonies, but later, along the more temperate middle reaches of the coast.

It was the French—spirited, aggressive, daring to the point of recklessness—who first braved the unknown lands and peoples of the interior. Frenchmen paddled upstream upon the continent's waterways, leading to vast inland seas, to headwater Lake Superior, and to the North Shore. That particular stanza of their great American woodland venture took the intrepid French adventurers almost 100 years.

To the wild North American shores and the Gulf of St. Lawrence in 1534 came Jacques Cartier, sailor and explorer, with his ships, captains, and crews. He visited only the Gulf this time. In the following year Cartier returned on another voyage and penetrated the continent by way of the St. Lawrence River for a distance of more than 500 miles. Cartier landed on the site of a future Canadian metropolis, spent the winter, and named the great headland *le Mont Royal*.

To Cartier in 1536 may be assigned credit for the name *Canada*—a name only slightly corrupted by the French from the natives' *Kanata*, which they called their own settlements—adopted officially for the entire country at the time of confederation, in 1867.

On a third trip, in 1541, Cartier established a post near the site of future Montreal. He traded with local Indians—and thus furthered development of the great fur enterprise that was to dominate the streams, lakes, and woodland portage paths of North America for three centuries. But for Cartier this was the last trip; he passed from the worldly stage in 1557.

After Cartier, other Frenchmen navigated the St. Lawrence, but little remains recorded of their penetration into interior North America for the next half-century. Then in the early 1600s came Samuel de Champlain with more ships and French adventurers.

Champlain—sailor-soldier, discoverer, geographer—coursed farther. He founded the city of Quebec, discovered Lake Champlain, visited Lakes Ontario and Huron, and—no doubt with the advice of local Indian travelers—sent his agents to Lake Superior.

Champlain did not follow directly up the train of the Great Lakes. But he and his company blazed a different water route to the upper lakes that was to be used as a standard in the next two centuries. His alternative, much shorter, left Lachine near Montreal and led west up the Ottawa River. The track followed up the Ottawa's tributary, the Mattawa River, across portages, through Lake Nipissing, down the French River to Georgian Bay on Lake Huron, and north and west to the sault, or rapids, that emitted from Lake Superior. Besides being shorter, this route had the added advantage of skirting the land of the troublesome Iroquois. Champlain, the Father of Canada, can as well be considered the "discoverer of the Great Lakes."

Sometime in the early 1600s, Étienne Brulé and companions reached the uppermost Great Lake, according to our best but scanty evidence—the first white men to view Lake Superior. Brulé had previously explored Lake Huron (1610) and Lake Ontario (1615). With information that Brulé and his agents brought back, Champlain thus constructed in 1632 the first map of the Great Lakes, including Lake Superior (but not Lake Erie and Lake Michigan). Although reliable details and the western end of the *Grand Lac* were lacking, the information probably having been obtained only as hearsay from the natives, the map clearly included the river and sault at the eastern end of the Superior basin—a detail that must have been tantalizing to later adventurers.

From this time of discovery, the rapids of the St. Marys River was to serve as the great meeting place of interior routes inviting

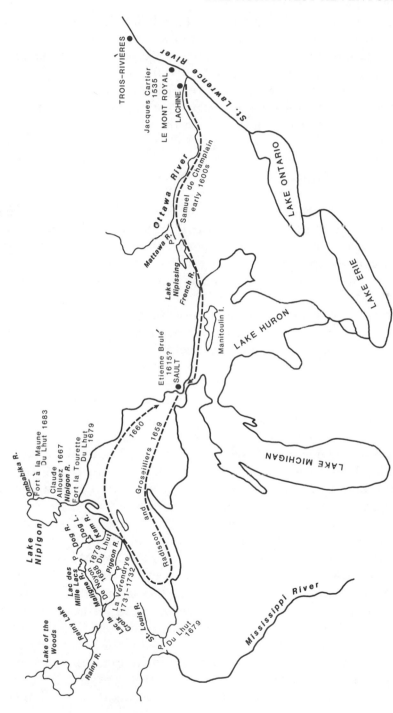

*The Great Lakes region and the routes and forts of early explorers in the North Shore area*

45

passage to the Northwest. It had served similarly in the past for prehistoric tribes and it was to serve for more in the future: trading post, mission, staging ground for western explorations, the great gateway through which all adventurers to Lake Superior and the North Shore must pass. And during the Second World War, the St. Marys River was to carry the heaviest ship canal traffic in the world, with iron ore from the North Shore.

The middle 1600s was a quiet period but also one of preparation. The colonies of New France were well established; trouble brewed with the Iroquois, but major calamities were beaten off. *Habitants* along the St. Lawrence River were high-spirited, adventurous, ever-ready to leave plow and family for a day's exploration, hunting, or trapping in the darkened woods behind their farms. In this ambience, the *voyageur*, soon to be typically North American, evolved.

Behind the spirit of adventure inherent in these sons of Normandy were three major motives of French colonization which were not always distinct: continuing exploration and the search for the Northwest Passage; the fur trade, the richness of which was not yet even imagined; and the saving of native souls.

For the pursuit of these motives, the stage was now clearly set. Exploration of the Ottawa River route to Lake Superior's outlet had been completed. Trading for furs along St. Lawrence shores was under way (and who could guess what wealth of beaver and other furbearers lay beyond the Sault?). And for the reluctant heathen souls, a growing army was forming for an all-out assault—Jesuit priests of the French Catholic Church's Society of Jesus, soon to be known as "black robes" to Lake Superior's Ojibway, Cree, and Dakota, as intent on scientific discovery and exploration as on soul saving.

Missionaries had already been busy on Lake Superior. Jesuits at the Sault in 1641 named their mission Ste. Marie du Sault, to honor the Virgin Mary, and thus effectively christened the rapids Sault Ste. Marie. A name for the river, St. Marys, naturally followed. Later in the same decade, the Jesuit *Relation*, the regular report of the priests' activities to their New France superiors, first mentioned the name *Supérieur* for the great inland sea. The assembled *Relations* remain among the greatest of historical documents.

Upon such a stage, carefully and painstakingly developed by

the French for more than a century now, stepped two new characters—more daring than most—to push the frontier of exploration farther west. These were Radisson and Des Groseilliers—two
of the greatest of Northwest adventurers. They were loyal Frenchmen, but their exploits in the Superior country were ultimately to
contribute to the fall of France in the New World—and to change
forever the political face of North America.

Médard Chouart was born in France, close by
the Marne River, in 1618. He was known to history by the title
derived from his family property, Sieur des Groseilliers. By 1645
he had arrived in the New World, at the city of Quebec, a young
man of twenty-seven years. His western adventures began almost
immediately.

Pierre Esprit Radisson was also born in France, perhaps in
1636, and was soon in the New World as a boy. His early adventures in North America included capture by the fearsome
Iroquois, while still a teenager, and a hair-raising tale of later escape and rescue. The two were known to each other; Radisson
was a half-brother to Des Groseilliers's second wife and thus recorded as Des Groseilliers's brother-in-law.

The place was Three Rivers, on the St. Lawrence, downriver
from Montreal, where plans for their Lake Superior voyage were
hatched. The year: 1659.

Three years before, Des Groseilliers had returned from a
journey to Lake Michigan and points west, where his tour of discovery had taken him possibly as far as the Mississippi River, but
not to Lake Superior. On that trip he had traded with the natives
and brought back a large quantity of furs, many canoe loads. As
rich a harvest as this was, it remained to him, apparently, merely
tantalizing.

For now, in 1659, Des Groseilliers was ready to go again, this
time with his brother-in-law Radisson, this time to Lake Superior.
Des Groseilliers, at forty-one years, was an experienced explorer
and trader; Radisson was but a boy of about twenty-three, yet a
toughened woodsman, both friend and fighter of Indians. Together, the partnership was destined for historic fame.

Radisson and Des Groseilliers left Three Rivers in August,
headed by way of the Ottawa River route for Lake Superior. They

departed with some haste and secrecy—apparently to avoid others who wished to accompany them—and without license from the governor of New France. These omissions, as it turned out, would earn them the great displeasure of New France administrators. It would also eventually bring about more far-reaching consequences.

The two partners arrived at the Sault and noted the Indian fishing for whitefish in the rapids (and, probably at the Indians' invitation, enjoyed the eating!) and then proceeded along Lake Superior's south shore in their canoes. At the southwestern end of the lake, Radisson and Des Groseilliers made friends with the natives, arbitrated disputes between tribes, built a cabin against the coming winter, and traded for furs. The winter passed profitably.

Next summer, the partners and Indian assistants struck out across Lake Superior, to the North Shore. It is unclear exactly where they landed—perhaps at a river mouth. History suggests that it was the Pigeon River, since this stream carried the name Groseilliers River on early maps. Later, apparently, the English translation of Groseilliers—Gooseberry—was moved down the map to a different stream, on which is now located one of Minnesota's most popular state parks.

Radisson and Des Groseilliers did not dally long but continued along Lake Superior's northern coast toward home. There is some speculation that on this trip they visited James Bay, through the Nipigon River-Albany River route, but the time element argues against it. From the Sault, they retraced the Ottawa River route to Montreal, and finally to Quebec.

They arrived laden with thousands of pelts, sixty canoes full, a wealth that was to constitute a large part of New France's revenue for the year, estimated at up to $300,000 in equivalents of the time. But now the partners—literally conquering heroes—were met by disappointment.

Because they had not obtained a license, and apparently because they had spurned some persons of influence who had desired to go with them, Radisson and Des Groseilliers were treated shabbily. Both men were arrested and fined; Des Groseilliers was thrown in jail; most of their wealth in furs was confiscated. The partners were embittered.

They felt they had accomplished an epic feat—which they had—but they were treated instead like criminals.

What did Radisson and Des Groseilliers really accomplish? What influence were these two to have on the country north of Superior?

Certainly, they had harvested part of a great natural resource and brought back wealth. Although exploitation of the new continent's resources had begun much earlier, this particular yield by the two partners opened up the eyes of the world to the immensity of the fur potential, previously unappreciated. This development created for France, and later England, nothing short of an economic revolution. Radisson and Des Groseilliers can be said to have initiated the *fur trade of the Northwest*, as we have come to understand and appreciate it in history.

The opening of trade in the Northwest was not the only impact of the two partners, however. They were explorers, and what they also accomplished on this trip was the opening of Lake Superior—the white European's first comprehensive view of the body of water so immense that early Indian accounts were interpreted as descriptions of the Pacific Ocean itself.

In the decade to follow, there was to be a flurry of activity on Lake Superior. While white explorers up to this time had mainly fur trading on their minds, the Jesuits in the 1660s began to make the first *systematic* exploration of the great lake. Jesuit missionary Claude Jean Alloucz, in a birchbark canoe, navigated around the western end in 1667 on his celebrated mission of mercy to displaced Nipissing Indians on Lake Nipigon. These travels contributed to the famous "Lake Superior Map," prepared by Allouez and Father Claude Dablon, published in the Jesuit *Relation* of 1672.

Radisson and Des Groseilliers had also made friendly and useful contacts with western natives. These contacts were to stand future travelers in very good stead, for the conduct of travel and business in the wilderness by the whites would have been totally impossible without the cooperation and assistance of the Indians.

Years later, Radisson was to put down in writing his and Des Groseilliers's adventures. Not as scrupulous a reporter as we might wish, Radisson nevertheless made a permanent, written record—the first such of human experience on Lake Superior and of European contact with the descendants of the region's early Americans.

But the greatest impact of the two explorers' trip was yet to come. For now, in 1660, disillusioned by treatment from their own government, frustrated by negative responses to their attempts to organize and carry out further trips, Radisson and Des Groseilliers turned to another government—England's.

Somewhere—perhaps from the Indians on the North Shore, perhaps from a Jesuit priest back in Quebec—Radisson and Des Groseilliers had heard of Hudson Bay as an easier and shorter route into the heart of the continent's beaver country. In his later narrative, Radisson suggests that the two actually traveled from the North Shore to James Bay, an unlikely event. But when English authorities and businessmen of London heard the two partners' story, they listened with attention. England, of course, had previously laid claim to the great saltwater bay in the middle of North America, the result of Henry Hudson's 1620 voyage.

So more voyages were organized to Hudson Bay, explorations were conducted, furs were obtained and brought to England. Radisson and Des Groseilliers, on separate ships, themselves weighed anchor in 1668, bound for the Bay. And, though Radisson's craft fell prey to storm, Des Groseilliers's expedition was successful—bringing back another fortune in beaver pelts.

The king's court and the businessmen were impressed—enough to organize a new company and to grant it extraordinary powers. The company received its charter in 1670, defining the organization as: "Governours and Adventurers trading to Hudsons baye." It was a sweeping grant that included trading license over vast territories, vaguely defined but almost infinitely inclusive, effectively granting hegemony over much of North America. Commercial enterprise of "The Bay" continues to this day.

The quaint spelling in the company's charter—"Hudsons baye"—left an interesting but annoying footnote in written history. The geographical body is *Hudson Bay*, whereas the institution itself became the *Hudson's Bay Company*—a minor typographical difference that has confused reporters and editors alike ever since.

The "adventurers" in the company's name were investors—

capitalists who were "adventuring" with their investments. But the army of pathfinders, fur traders, missionaries, governors, voyageurs—those who followed the misty paths of Radisson and Des Groseilliers into the Superior country—these were adventurers indeed.

Alarmed at increasing British influence on the edges of its New World dominions—especially with England's formation of the Hudson's Bay Company—France responded with a stirring act of territorialism. In a colorful ceremony beside the river at Sault Ste. Marie, in the spring of 1671, with noblemen, soldiers, Indians, and missionaries in attendance, French representatives in traditional style officially took possession of what may be fairly described as all of interior North America. The scene was accompanied by waving swords and crucifixes, by the sounds of musketry and prayers.

The claim included, of course, Lake Superior, its tributary waterways, and all surrounding lands. The claim also embraced Hudson Bay, an inclusion that later was to cause bitter and sometimes violent contests with the English—who had previously claimed the bay—on those northern shores.

François Daumont, Sieur de Saint Lusson, French nobleman, laid France's claim in the name of Louis XIV. The effusiveness of the pageant was intended to convey not only the immensity of the claim, but also—we may suppose, especially, for the benefit of attending Indians—its legality.

In the growing French colonies along the St. Lawrence River, however, a strange and contradictory dichotomy of French purpose set in during the decade of the 1670s. For, despite an appreciation of the potential fur harvest that Radisson and Des Groseilliers had engendered, the stimulation from rival England in establishing the Hudson's Bay Company, and the claim to territory at the Sault—the government policy of New France now attempted to discourage further interest in interior exploration and trade. Wishing to stabilize New France in the form of any other French province, that is, with a system of agriculture around urban trade centers, the royal government in the faraway mother country failed to appreciate the fact that its distant citizens were not in a pastoral French province but on the edge of a wilderness. Vast, rich, unexplored—that wilderness beckoned with adventure and invited commercial exploitation.

A royal decree in 1673 prohibited persons from leaving their habitations to wander in the woods for more than twenty-four hours (upon pain of death!), and in 1676 another law prohibited travel to trade with the Indians. Given the high-spirited character of the French colonists in this geographical setting, the result should have been predictable: the king's decrees were ignored.

In this environment, then, came the full development of the *coureurs des bois* (runners of the woods), independent individuals who departed the confines of city and village to explore and trade as self-sufficient entrepreneurs. They were frequently classed as outlaws, but they flagrantly left the settlements by the hundreds. The king's representatives in New France could do little but cringe and complain.

Onto the stage of this dynamic state of affairs now stepped another giant of a figure. Strong-willed, of noble parentage, and a royal soldiering son of France, Daniel Greysolon arrived in the colonies in 1672, to serve in Quebec as captain of marines. He is better known to history by his landed title, Sieur du Lhut—destined to become one of the great characters in the drama of exploration on Lake Superior, namesake of the future Minnesota city and international port at the western end of the lake.

After returning to France to participate in military campaigns against the Dutch, Du Lhut was presently back in North America. In the manner of the *coureur des bois*, he made plans for western discovery.

In the early fall of 1678 Du Lhut left Montreal with his brother, Claude Greysolon, Sieur de la Tourette, and a company of Indians and other Frenchmen. Like many others, including Radisson and Des Groseilliers nearly twenty-five years before him, Du Lhut failed to obtain approval from the authorities. The omission was later to cause him grief, as well as apparent qualms of conscience. But for now, in the invigorating crispness of a northern autumn, and with Lake Superior's waters beckoning from far to the west, Du Lhut's conscience must have been suppressed. He was forty years old, experienced, tough, confident.

Du Lhut and his party took the now-familiar Ottawa River route and arrived at the Sault later in the fall, to spend the winter along the rapids of Lake Superior's outlet. In the following spring, the party put out again, across the big lake, along the North Shore.

One of the purposes of the Greysolon brothers' expedition was to interdict Indians trading from Lake Superior to Hudson Bay. First, a post was built near the mouth of the Nipigon River—Fort Nipigon, or Fort la Tourette. Continuing westward, the brothers erected a second station, at the mouth of the Kaministiquia. This palisaded post was the first European establishment on the northern coast of Lake Superior, the first of several at this location, and a forerunner of Fort William, great depot, successively, of the North West Company and Hudson's Bay Company, more than a century in the future.

Continuing west, Du Lhut's party landed weeks later, in June of 1679, on the big sandbar at the lake's westernmost end, stepped ashore to make their landfall and greet the native inhabitants. The site was Little Portage, at the base of Minnesota Point, a historic location later to become the Duluth Ship Canal.

Sieur du Lhut continued westward overland and explored, in part, the St. Louis-Savanna Portage leading toward the Mississippi, as well as the Brule-St. Croix rivers route from Lake Superior to the lower Mississippi.

On later trips to Lake Superior, Du Lhut was to build several more forts—trading posts actually, rather than military establishments—at critical locations that were forerunners of a great network important in the fur trade.

The latter part of 1683 found Du Lhut in the Nipigon country again, and in the next year he constructed Fort á la Maune, near the mouth of the Ombabika River on Lake Nipigon; this post—later commanded by his brother, La Tourette—was also sometimes called Fort la Tourette. More posts followed in the Nipigon country and along the Nipigon-Albany route to Hudson Bay, again, to disrupt the English Hudson Bay trade. Du Lhut kept an interest in Fort Kaministiquia for about a decade, using it as a base of operations for several trips, until about 1688.

After that date, Du Lhut's activity on the North Shore tapered toward a close. Back on the Ottawa River, he commanded a force against the Iroquois that saved Montreal from possible disaster. But the great explorer of Lake Superior and the upper Mississippi, arbiter of Indian troubles, and builder of trail-blazing posts on the North Shore, died soon after. The old soldier was laid to rest in a church graveyard in Montreal in 1710.

An inevitable decline in French influence began in the latter part of the seventeenth century. The early explorers had done their job well—discovery, mapping, establishment of posts and missions, quieting of Indian conflicts which threatened trade. Above all, the French explorers had opened the eyes of the world to the potential harvest of fur from this Northwoods wilderness—and to the fortunes to be garnered.

But it seemed that the French pathfinders had done their job too well, for an overabundance of furs resulted. The markets of France had not yet developed the capacity to absorb the quantities of fur imported from its colony's enterprise. In 1696, Louis XIV revoked all fur trade licenses because of the overabundance, opening the door for British intercession in the fur trade in the Great Lakes country.

The breach between France and its North American colonies was further opened in the early years of the 1700s by the War of Spanish Succession, actually between France and England. In 1713, by one of the Treaties of Utrecht that closed the war, France surrendered much of the claim it had made in the 1671 pageant at the Sault: all interests in the Hudson Bay country, as well as Newfoundland and Nova Scotia.

The halcyon days of French influence and the great French explorers were nearly over. The era of French domination of middle North America had been concurrent with the age of discovery in the Superior country. That age was about to give way— from exploration to industry, from French to English.

One of the most important institutions in the fur trade in the land of Lake Superior was to be the North West Company's great establishment of Fort William, near the mouth of the Kaministiquia River early in the nineteenth century. Back in the late 1600s, however, its first predecessor had been founded by Sieur du Lhut, whose post depended greatly upon the military and political events unfolding across North America over a period of more than a century.

Du Lhut must have had some inkling of the future significance of the Kaministiquia River site when he first visited it in 1679. With English activity on Hudson Bay now completing its first decade and escalating, the importance of this stream leading northward from Lake Superior could not have been lost on the astute explorer.

At the close of Du Lhut's decade of exploration and periodic use of the Kaministiquia posts, we find Jacques de Noyon, canoeman, possibly an employee of Du Lhut, on the Kaministiquia River. De Noyon must have been fascinated by this large stream, emitting from unknown sources in the interior. In 1688 De Noyon investigated the river as a canoe route, met the great obstacle of Kakabeka Falls, and discovered the way around it to found Mountain Portage. Upstream on the Kaministiquia, De Noyon blazed an epic trail through Dog Lake and up the Dog River, then westerly overland to Lac des Mille Lacs, to Lac la Croix and Rainy Lake, a heroic venture. We will return to this historic canoe route, of great significance in future fur trading, in the next chapter.

But for now, the major interest of the French in this region was still to divert Indian trade from Hudson Bay, and posts on the Nipigon route were much better located for this purpose. Furthermore, French interest in the Northwest hinterlands was soon to be diverted to the Pigeon River, a much easier route than the Kaministiquia. And so the Kaministiquia River post declined in importance. Off and on, French commanders arrived to govern it, but when France lost North America to England in the French and Indian War, the post was lost as well.

At this point, the Kaministiquia location fell into obscurity; the British, even as new owners, passed it by for another forty years. Like the Kaministiquia River canoe route, Fort William will be discussed in more detail in the next chapter.


Even though the eras of early exploration and the fur trade were inextricably linked and always overlapped to some degree—for the entrepreneurial interest in fur was one of the major incentives for exploration—the immense potential for fur harvest was not yet being fully exploited by 1700. One of the major reasons for the delay was the lack of exploration of the interior north and west of Lake Superior—the great Northwest—fabulously rich with beaver and other fur-bearing animals, and yet virtually unknown.

One more Frenchman was to step into this scene—the last of the great Northwest adventurers. In the late summer of 1731, he stepped onto the shores of a small bay, not far from the mouth of the Pigeon River, to blaze an epic trail of canoe routes and posts into the interior.

Born in Three Rivers, New France, in 1685, Pierre Gaultier de Varennes was a true son of America, the only one of the major French explorers to claim the North American colony as his birthplace. Like others, Pierre de Varennes in his exploits in the Superior country is known to us under his title, Sieur de la Vérendrye. He was to be the last, some say the greatest, of the famous French adventurers searching for *la gloire* in the Northwest and along the Superior North Shore.

Son of the governor of the Three Rivers district, La Vérendrye nevertheless went to Europe as a young man to fight in France's army, was wounded, and returned to the St. Lawrence colony as an officer in a colonial regiment. Married, he sired four sons. Fired with plans for exploration, his sons grown, La Vérendrye at last left for the Superior country.

La Vérendrye had wealth enough to support his initial ventures of western discovery, but not enough to pursue single-mindedly an ultimate goal, the discovery of the Sea of the West, an imagined saltwater sea of the interior, leading to the Pacific. The king, Louis XV, gave him official sanction for fur trading but no financial backing. Consequently, La Vérendrye acquired additional financial support from Montreal and Quebec merchants interested in opening up the trade in furs. La Vérendrye was constantly balancing his efforts between his noble quest for the western sea and the commercial need for the development of the fur resource—the latter always to support the former. Although he never found the Sea of the West, history has granted him preeminent success in both exploration and commerce.

By 1726, Sieur de la Vérendrye was on Lake Nipigon, commanding Fort á la Maune at the mouth of the Ombabika River, set up years ago by Sieur du Lhut. And here it was that his imagination was fired by native tales and a crude map, sketched on a scrap of birch bark by an Ojibway Indian. The map showed a remarkable route to the interior Northwest—inland from Lake Superior by the Pigeon River, through a westward-leading bead-chain of lakes toward the Sea of the West. It was a treasure map indeed.

In August of 1731, La Vérendrye, with three of his sons, a nephew, his Indian cartographer, and a supporting company of about fifty others, landed on Grand Portage Bay, birchbark map in hand. Although problems with his company forced La Vérendrye to delay his own travel farther, one son, a nephew, and several others of the party penetrated inland. The group used a trail known by the Indians as the "Great Carrying Place", a nine-mile portage that circumvented an impassable twenty-mile reach of the Pigeon River filled with wild cascades and waterfalls. The trail leading inland from the North Shore was the route that was to be the most famous of all in the fur trade, the Grand Portage.

Farther inland, the route continued up the Pigeon River and through the bead-chain of rock-bound lakes to Rainy Lake. Part of this route, from Lac la Croix to Rainy, had already been traveled by Jacques de Noyon in 1688. And no doubt the Ojibway had worked out, long ago, the intricate waterways.

On Rainy Lake the La Vérendrye party built a post in late 1731—Fort St. Pierre. This early post and its successors at the same location continued in operation until the late 1800s. On this famous water trail from Lake Superior westward, the Rainy Lake post was the first of many.

The following year, La Vérendrye and party made a second trip to Grand Portage Bay. Over the nine-mile portage, they penetrated to the post on Rainy Lake and beyond, to Lake of the Woods. Here a second post, Fort St. Charles, was constructed. La Vérendrye continued to found posts on Lake Winnipeg, Red River, and others.

In a century to follow, other explorers and traders were to establish in the interior a vast, intricate network of forts and trading posts. Their empire of posts extended north and west from Lake Superior throughout all of northwest North America, from the North Shore's Pigeon River to western Canadian prairies, north to Athabasca, to the Rocky Mountains, to the Pacific Ocean.

Following in the steps of La Vérendrye on the Grand Portage, and in his wake on a chain of lakes that was to become part of the border between the United States and Canada, came the great army of the North American fur trade—traders, voyageurs, Indians, company administrators—to gather enormous fortunes in fur.

History does not award sole credit to La Vérendrye for discovery of the Grand Portage canoe trail. De Noyon had preceded him to Rainy Lake, by the Kaministiquia route, and obviously the natives were thoroughly familiar with the interior. But at a time when a breakthrough to the interior was needed, La Vérendrye and his sons and nephew recognized the significance, saw the opportunity, and grasped it. Well supported by equipment, supplies, and men of expertise, La Vérendrye exploited the opportunity. True, his principal goal of finding the Sea of the West was unattainable; the quest for a westward-leading pathway to interior North America, however, was a resounding success.

In subsequent years the La Vérendrye company went on in their search for the western sea—to the beaver meadows of the Saskatchewan River country, to the plains on the upper Missouri, the Black Hills, possibly to view the distant Rockies. By sailing ship, canoe, moccasin, and horse, La Vérendrye and his sons blazed the last of the great Northwest trails of discovery.

The roll call of explorers on the shores of Lake Superior is larger than recorded history would have us believe. The prominent figures we know about had companions, assistants, canoemen, and guides Indian and white—men whose names were not recorded. Others forged trails into unknown regions but did not tarry long enough at the time, nor take time in later years, to fill out a journal or pen a manuscript; but their word-of-mouth instructions and personal guide services were nevertheless invaluable.

Still others entered the interior but did not return; only a few names of those lost souls were recorded. No doubt, some of them were named Jean Baptiste, Jacques, and François, first names that appear repeatedly throughout history's record of the French period. A few of the well known early explorers—Pierre Charles Le Sueur, Nicolas Perrot, Sieurde La Salle, Louis Jolliet—

visited the North Shore only briefly or had only peripheral interests there.

Many missionaries of the Society of Jesus, the Jesuit priests, conducted their work of religious conversion and education in relative obscurity. In addition to Claude Allouez, we find in their *Relations* other explorer-priests named in the Superior country—Claude Dablon, Pére Jacques Marquette, Gabriel Dreuillettes—but most remain lost in the mists of history. The "black robes" were as assiduous in their pursuit of science and geography as in religion, if not more so.

After La Vérendrye opened up the western Superior country, the French dominated the hinterlands—but only for another forty years.

During this time, the English seemed content to sit secure at their Hudson Bay posts and to trade for pelts that Indian middlemen would bring to them from the interior. Lacking canoe-building material because of the absence of birch trees, without a colonial labor force or even the expertise for canoe making, and with their headquarters far across the ocean, it seemed a good system for the Englishmen. The enterprising Frenchmen, with more favorable logistics, continued to build posts at strategic points on rivers and lakes, intercepted the Indians and their pelt-laden canoes on the way toward Hudson Bay, and captured the trade. And at Grand Portage the French exploited the Pigeon River-Rainy Lake route with great alacrity, as they once had done at the old post at the mouth of the Kaministiquia River.

But in the 1750s troubles in Europe erupted into the Seven Years War between France and England, to spill over into North America as the French and Indian War. French explorers and traders, many of them militia officers, were called to duty in New France to fight the British. Use of Grand Portage and the interior routes declined.

The tides of war did not favor the French. And even the fact that the Ojibway sided with them did not swing the balance. In 1759 General Louis Joseph de Montcalm and the French lost to British commander James Wolfe on the Plains of Abraham, at Quebec, the war's critical battle. The year 1763 brought an end to the war—and to French claims and colonies in the New World.

France's defeat also brought a close to its posts in the country north of Lake Superior—and as well to its canoe brigades venturing to find the Sea of the West and the Northwest Passage.

The end of the French-English conflict marked the end of the halcyon days of exploration and yet a beginning. The struggle for discovery, now successful, was spent, and the rush for fortunes in fur was now to soar toward its zenith. French colonies along the St. Lawrence had produced a great labor force uniquely qualified to serve in future canoe brigades, the voyageurs.

The North Shore's tributary waterways had served well the trail-blazing French explorers; the interior trails were now to serve as well the British legions that were to follow, in their quest for *Castor canadensis*.

# FOUR

# *Empires of Fur Posts*

I n the wake of war and its dislocation of governments and institutions came a motley collection of new adventurers to the Great Lakes country. Some were English, some British colonials, eager to assume the responsibility for the administration of the new land—and the profits—from which the French had been forced. Other newcomers came from abroad, or were left over from the war, like the Highland Scots. And some Frenchmen stayed. All were drawn by a siren song from western forests, lakes, and streams: the fur trade.

Earlier in the French regime, a certain spirit of independence had developed in response to the siren song. French farmers were stirred by the western exploits of heroic figures in the age of exploration. That spirit grew in the *coureurs de bois* (runners of the woods, bush rangers). By deserting home, farm, and church to roam the western forests, the woods runner was branded an outlaw by some. Many were arrested; one was publicly hanged. But to simply fix their spirit as irresponsibility was gross oversimplification, for the spirit of independence was combined with entrepreneurism, adventure with risk.

A small fur trading business could get a start by two coureurs de bois with a canoe and a load of trade goods. The French colonial government licensed such trading ventures and

61

limited the number of licenses issued. But to a foot-loose French-man along the St. Lawrence, eager for adventure and profit in the interior, the lack of a proper license was little deterrent. Often capital was obtained from a Montreal merchant who winked at the breach of rules—and who might share in anticipated profits.

By 1763, the runners of the woods had become legend. Their spirit of adventure and free enterprise—ripening for more than a century—was now to be the very pulse and tenor of the Northwest fur trade. That spirit did not leave with French authorities who departed across the ocean, for it was now a spirit of North America, and it was merely to be inherited by new proprietors.

The British were slow to start, after they had won Lake Superior. For the first four years following the close of the French and Indian War, the government in London restricted trade. Only the larger, established posts were operated, and in the Superior region Fort Michilimackinac, on the south side of the Straits of Mackinac, was the single major station. In 1765, only one trader—British colonial Alexander Henry, from the Mich-ilimackinac post—was given license for trade on Lake Superior.

There were good reasons for the Englishmen's hesitancy. The French had built up a great logistic infrastructure of posts, supply routes, a labor force of voyageurs, knowledge in the ways of wilderness survival, skill in birchbark canoe building, and rapport with the native Ojibway. The fact was that French moccasins had worn well the portage paths north of Superior for a century—and they were now hard to fill.

Of course the British knew the river trails to Lake Superior from James Bay, principally the Michipicoten and Nipigon rivers. But an English network of inland communication—from either Lake Superior or Hudson Bay to the interior Northwest—was still to be developed.

The vast expanse of North American forests was a wilderness only to the Europeans. To the native Ojibway and Cree around Lake Superior, the land was a known environment, its far-flung landmarks articulated by thousands of navigable lakes and streams. The routes were traveled by the native American in

*Beaver Lodge*

a vessel so admirably suited to these waters that its design has survived, with no fundamental modification, to the present day.

The north canoe *(canot du nord)*, constructed of birch bark that was available along almost any nearby stream bank or lake shore, was endemic to North America. The canoe had evolved through its long use by the early Americans on the continent's unique system of water routes. Upon it now depended the success of the fur trade, and the Europeans readily adopted it.

For two centuries—about 1615 to 1821—Lake Superior became the great crossroads for these routes of exploration and trade to the interior Northwest. To Lake Superior ran at first an all-French highway—from the St. Lawrence River and Montreal to the Sault at Lake Superior's outlet. During this period, the English dominated the other major route into the continent— Hudson Bay—and later Lake Superior, too.

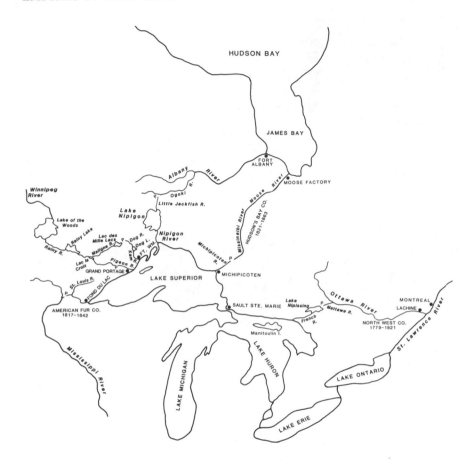

*Major routes of the North West Company, Hudson's Bay Company, and American Fur Company, and their principal fur trading posts, in the North Shore area. P = location of portage between major drainage basins.*

From Superior's North Shore, major lines of water communication were pushed to frontiers north and west. Two tributaries—the Kaministiquia and Pigeon rivers—constituted the main routes to Rainy Lake, Lake Winnipeg, and then beyond. The St. Louis River, at the extreme western end of the big lake, led north and joined the Kaministiquia-Pigeon route to Rainy Lake. The St. Louis route in addition connected to the Mississippi River by portage and the Savanna Rivers. Two routes connected Lake Superior with Hudson Bay by way of North Shore tributaries—the Michipicoten and Nipigon rivers.

In the heyday of the fur trade, the Superior country was no wilderness of solitude. The river trails often bustled with the activity of birchbark canoes.

After France's claims in North America were extinguished, British individuals and the North West Company traded over the same North Shore routes. In 1821, the two great fur trading companies—North West and Hudson's Bay—were joined in an amalgamation, retaining the Hudson's Bay Company name. And since Hudson Bay was a shorter and less costly route into interior North America, Lake Superior and the North Shore tributaries declined rapidly in importance and use.

Little remains today to mark these North Shore river routes of discovery and trade. Agriculture and settlement were anathema to the fur trade and were not encouraged; so villages and towns did not result from inland posts. Only along the coast have major depots led to settlements and later cities. To the once busy inland streams and lakes, the solitude of wilderness returned. And today, where the old canoe trails are crossed by modern highways, only historical plaques mark their locations.

Shortly after England's initial prohibition was lifted, British trader John Erskine in 1768 hacked a clearing from the forest around a small embayment in Lake Superior's coast, not far down the lake from the mouth of the Pigeon River. Probably he built a primitive post there, too. It was the beginning of what was to become the most important entrepôt of the fur trade in its time—Grand Portage.

The little bay was an ideal location, in several ways. The surrounding ground was flat—a rarity along this part of the coast—with room for building construction and a gravel beach for the easy landing of canoes. The bay itself was protected from northwest storms by almost mountainous terrain in back of the clearing, from the coast's nor'easters by a lava headland forming the east side, and from the fury of Lake Superior's storm-driven seas by Grand Portage Island, in the mouth of the bay.

But the main reason for selecting this site was its proximity to the Pigeon River, ten miles to the northeast. This river led to the bead-chain of lakes sketched by La Vérendrye's Ojibway cartographer on his birchbark map, an ancient Indian track through a vast wilderness otherwise unfathomable. The lower reach of the Pigeon was a tortuous, falls-and-rapids stretch impassable to river traffic, dropping nearly 700 feet in elevation in twenty miles of length, including obstacles today known as The Cascades, Horne Rapids, Middle Falls, and culminating in 100-foot High Falls.

An old native portage around this wild reach of river circumvented the waterfalls and cascades. The trail was nine miles long, rough, muddy in places, and mostly uphill from Lake Superior, where glacier-worn gaps in the mountainous terrain made the task at least possible, if tough: the Ojibway's Great Carrying Place. At the inland terminus of the trail, the portage ended in a spruce-covered river landing that was to become Fort Charlotte, staging depot for the inland trade.

One more waterfall, seventy-foot-high Partridge Falls, three miles up from Fort Charlotte, had to be portaged. But from this point on the river was mostly flat, leading west by a series of rock- and pine-bound lakes—first upstream in the headwaters of the Pigeon, then downstream in the headwaters of the Rainy. This track was the line of water-communication that was later to constitute the border lakes in the center of the Quetico-Superior wilderness, part of the boundary between Canada and the United States.

Over the Great Carrying Place had tramped prehistoric early Americans and then La Vérendrye's company of explorers. In the years immediately following his historic arrival at Grand Portage Bay, La Vérendrye and his sons built up an encampment, with dwellings, blacksmith shop, and warehouse. Now, in 1768, following John Erskine's clearing and post, the trail was to receive the tramping of many thousands more, British traders and French-Canadian voyageurs. They were to remove incalculable numbers of pelts of beaver and other furbearers.

At first, the traders' partnerships were small, independent business arrangements. In the interior, wintering partners built modest, temporary posts that were not intended to last for more than a year or two. Winterers frequently comprised one Britisher and a Frenchman left over from prewar times—the latter being very useful with his knowledge of the land and the ways of Indian custom and language. Alexander Henry from Fort Michilimackinac teamed up with Jean Baptiste Cadotte, who broke many a trail for the famous British trader.

When small groups of traders expanded, leaders emerged and voyageurs were engaged. The role of Montreal merchants evolved from simple sponsors to strong and influential partners themselves. These were the Montreal partners, who from their eastern offices supervised administrative affairs. The wintering partners, those adventurous associates who spent the snow-filled months at remote posts, traded with the Indians.

These partnership arrangements of free-traders continued to be independent—and individualistic. Competition for trade with the natives was inevitable, and conflict—even violence— arose on remote forested trails and at isolated posts. Eventually all concerned realized that the undisciplined arrangements were wasteful and unprofitable; some greater organization seemed necessary.

Thus, in the mid-1770s a group of Montreal traders, recognizing the need for cooperation, finally came to terms. About 1779, this group formed the North West Company, the organization that was destined to lead the Lake Superior fur trade for more than forty years.

Grand Portage became the North West Company's central field headquarters. From Montreal on the St. Lawrence, through Lake Superior, and into the Northwest to its Fort Chipewyan on Lake Athabasca, this British company operated Canada's first main throughway of commercial communication—3,000 miles. Its system evolved to perfection.

The first part of the water highway—Montreal to Grand Portage—was an arduous route of large rivers and broad inland seas. The other part—Grand Portage to Athabasca—was a maze of intricate inland waterways, small rivers and lakes, rushing rapids, punishing portages. The overall objective was to exchange trade goods from Europe for pelts from the natives—and to do it all in one season. A scant five-month period was ice-free. The round trip—6,000 miles, with the two very different types of waters to navigate, requiring different types of watercraft—in the short time available, was clearly impossible.

Using two types of vessels, both birchbark canoes, leaving either end of the route at the same time, in spring, made it possible to make the exchange in the available time. The two groups met at Grand Portage in midsummer; and both groups returned to their place of origin by ice-up in the fall.

From the mouth of the Ottawa River, the Montreal partners embarked in May in their large Montreal canoes loaded with trade goods, powered by voyageurs. The Montreal canoe carried a load of three tons—brass kettles, knives, needles, woolen cloth and blankets, guns and traps, and brandy—all the trappings of European civilization upon which the North American natives became dependent.

On Lake Superior, storms, wind, and waves took their toll. On portages, four men carried the canoe; others carried trade goods in ninety-pound packs, two at a time, three trips, for a total of 540 pounds. Horses and oxen were tried—unsuccessfully—on the Grand Portage.

These brigades took the old explorers' route up the Ottawa River toward Georgian Bay on Lake Huron. On big water, the Montreal traders coasted west along the north shore of Georgian Bay and North Channel to the Sault, then nearly another 500 miles along the North Shore of Lake Superior to Grand Portage Bay. The trip required eight weeks.

When the brigades rounded Hat Point on the east side of Grand Portage Bay, it was a time of rejoicing, song, and the flutter of the voyageurs' brightest red sashes.

In May, also, as ice broke in the north country, other brigades left for Grand Portage. These were the wintering partners and their voyageurs who had spent the long winter of seven months in remote posts, trading with the Indians for furs. They embarked in the *canot du nord*, carrying a load of one and a half tons.

Brigades were organized into four to eight canoes each. Cargoes were mainly of beaver pelts for the hats of gentlemen and soldiers, and also fine furs of ermine, lynx, and marten for special fashions in Europe. Landings were made at Fort Charlotte, at the northwest end of the Grand Portage, and the ninety-pound packs of fur were carried down the nine-mile trail to the great meeting.

After seven months of winter isolation and hundreds of miles of toil on portage and river, the meeting at Grand Portage with old friends from Montreal, rest and festival, reveling and dancing, must have been joy indeed. The season: mid-July.

The stockaded post, with its canoe yards, warehouse, and Great Hall, and sweeping grounds, was the western headquarters of the North West Company. Each summer, only two weeks' time served for the partners' business and the voyageurs' revelry. The rendezvous was the very linchpin upon which the continent's first great commercial enterprise pivoted.

But the two weeks were soon gone, business completed, appetite for festival satiated. For each group, the next two months were to consist of more toil and danger, on their respective return trips. A good season meant fortunes for the partners upon their return to Montreal.

Meanwhile, another scene of great import was unfolding to the south and east, an event that was to have the most profound effect upon Grand Portage and the North West Company: the American Revolutionary War.

Although Grand Portage lay upon ground that could have been hotly contested between American and British interests, no military skirmishes took place. In 1778, the British military establishment dispatched from Fort Michilimackinac a small detachment of soldiers to Grand Portage "for the purpose of preserving order," a minor act of apparent territorialism. By summer's end the soldiers left, ending the only military "action" of the Revolutionary War on the North Shore.

But the war's close brought a new political entity to Grand Portage. The Treaty of Paris in 1783 specified a border that was to follow the old line of the voyageurs' track, westward from Lake Superior to Lake of the Woods. But no one was quite sure at the time exactly which line of "water communication" the new boundary would follow, or whether Grand Portage was on British or American soil. So business by the British continued as usual, for more than a decade, at the great entrepôt.

In these uncertain times, competition among the North West partners became more intense, and for a while rivalries boiled over. A rebel group temporarily withdrew from the company in 1797, formed the "New North West Company"—popularly known as the "XY Company"—and began erecting its own posts, including one at Grand Portage. Violence flared on Grand Portage Bay. When at last cooler heads prevailed, the prodigals returned.

In 1798, forty miles up the shore from Grand Portage,

North West partner Roderick Mackenzie, in preparation for a possible move, "rediscovered" the old Kaministiquia River route of explorer Jacques de Noyon.

Eventually, around the turn of the century, American customs officers, flexing new sovereign muscle, showed up at Grand Portage and threatened to impose taxes on the traffic in trade goods and peltries. The North West Company yielded. Between 1801 and 1803, it constructed a new fort, soon to be known as Fort William, near the mouth of the Kaministiquia.

The Grand Portage post fell into disuse and decay.


The Kaministiquia River, just before it enters Thunder Bay on the North Shore, breaks into three branches, or distributaries—a common arrangement when a large river enters a receiving body of water across its own delta. Early forts in the French period, beginning with Du Lhut's first Fort Kaministiquia, were built on two of these branches, today termed Mission and McKellar rivers.

Now in the first year of the nineteenth century, as American threats of customs duties spurred the North West Company to yield Grand Portage, the company constructed their new post on the north bank of the main Kaministiquia River, the northernmost branch, one-half mile upstream from the mouth. The first rendezvous was held in 1803. The new establishment, larger and more resplendent then the Grand Portage depot, was first called New Fort and rechristened Fort William, for leading partner William McGillivray, in 1807. In that year the new post became the communications center of the transcontinental fur trade.

Fort William resumed the role that Grand Portage had previously played, yet on a grander scale. It became a huge complex with many buildings surrounding a large open square, including the Great Hall, many residences, council house, storehouses, a jail. The whole complex resembled a bustling town. Many shops and artisans of all trades were necessary, even in off-seasons when the partners and voyageurs were absent, to maintain buildings, to build canoes, barrels, iron tools, tin goods, and a host of other necessaries. Outside a sturdy stockade were shipyards for constructing Lake Superior vessels, and farm fields and livestock. It was a far cry from the crude stockade and log buildings of old

70

Grand Portage. Fort William became the nerve center and supply depot for the entire Northwest.

Summer was again the time for rendezvous, when the brigades of Montreal canoes arrived from the east, loaded with trade goods, to join with western brigades of *canots du nord*, loaded with fur. At such times, the population of Fort William perhaps numbered 2,000. The company's affairs of state were settled in the council house, and afterward the old rites of festival prevailed.

Absent now was the labor of transporting trade goods and fur over the grueling nine-mile Grand Portage. But instead was a more difficult and longer canoe route west, pioneered by Jacques de Noyon, in 1688. The first major obstacle, some thirty miles upstream, was the formidable, 128-foot Kakabeka Falls, around which climbed Mountain Portage.

Above the falls, the route led upstream on the Kaministiquia, to Dog Lake and then westward by the Savanne Portage across the continental divide. To large and complex Lac de Mille Lacs, the route continued southwest and, by way of the Maligne River to Lac la Croix, it joined the route from the Pigeon River. Eighty miles longer than the route from Grand Portage, the "Kam-Dog" route nevertheless served as the vital connecting link in the booming continental trade.

Still, for all its enriching and expanding enterprise, Fort William's usefulness was destined to be short-lived. Even while business continued with great success, other events beyond the company's control were proceeding little noticed, to both north and south; these events were to have debilitating effects upon the company.

From below the new border, John Jacob Astor's American Fur Company, newly formed in 1808, posed a competitive threat. Astor occupied the site at Grand Portage and had ambitions in the fur trade from Lake Superior to the Pacific Ocean. No longer would the Nor'Westers rule the fur trade in the Lake Superior country.

And when the United States, having run out of patience with British naval practices, declared war in 1812, shipping on the Great Lakes was essentially halted, including the orderly progression of Montreal canoe brigades. The North West Company diverted its energies to the conflict on the side of the British, who attacked and captured American Fort Michilimackinac. United States forces, in retaliation, attacked and burned the North West Company post and facilities at Sault Ste. Marie. But the year 1814 brought an end to the war, and both sides attempted to return to normal business in the fur trade.

To the north other events were brewing that would have more serious effects upon the future of Fort William and the North West Company. The Hudson's Bay Company, ponderous and slow, dignified and tightly disciplined, was eventually to be overwhelming. Competition in the trade had always been keen, even violent; the North West Company had the great disadvantage of the long, expensive inland route from the St. Lawrence, whereas the London adventurers enjoyed their much shorter route by sailing ship to the shores of Hudson Bay, fully as far west into central North America as Fort William itself.

This competition was bad enough—some North West Company partners had already suggested a combination of the two companies—but now Hudson's Bay presented an intensely feared threat to traditional conduct of the fur trade: agricultural settlement.

By 1811, Scottish nobleman Thomas Douglas, Fifth Earl of Selkirk and a large shareholder in the Hudson's Bay Company, had obtained an enormous land grant of 116,000 square miles on the upper Red River, including the area of present Winnipeg. His plan was to bring over from Scotland small farmers who had been dispossessed of their lands, establish them on the fertile prairie lands along the Red River, and, under the aegis of paternal Hudson's Bay Company, form a settled colony.

It was a noble experiment. But agricultural settlement had never been reconcilable with the fur trade. The company feared the loss of the buffalo, upon which the western brigades depended for the production of pemmican, a staple of dried buffalo meat.

In June of 1816, at a point in today's city of Winnipeg then known as Seven Oaks, fighting erupted that led to the killing of twenty-one settlers and the colony's governor. Surviving colonists were sent packing.

Enraged at the news of the "massacre at Seven Oaks" and believing that the Nor'Westers were behind it, Lord Selkirk with an armed company marched on Fort William. Confrontation followed. Selkirk arrested fifteen North West Company partners, sold off company properties, and occupied the great depot for the better part of a year. The North West Company's fur trading operations on the North Shore were paralyzed.

Legal battles between the two companies followed in England, and eventually the Nor'Westers were returned to Fort William. But litigation continued unresolved, for many years, further consuming the vitality of both companies. Definitive blame for the Seven Oaks tragedy was never fully established.

By now, as nearby beaver populations declined, the frontier of profitable trapping had been pushed farther and farther westward, resulting in greater expense and slimmer margins of profit. And markets for fur in Europe were failing.

Weary of debilitating and costly competition, climaxed by the violence at the Red River colony and the enervating litigation that followed, the two companies were forced into negotiation and, ultimately, cooperation. Lord Selkirk, still bitter over the Seven Oaks affair, opposed union of the two companies throughout the rest of his lifetime. But in 1821, one year after Selkirk's death, amalgamation into a single institution became a reality. It took the name of the London-based company: Hudson's Bay.

The North West Company, Fort William, and the old Montreal-Ottawa River route to Lake Superior fell for a long time into the dusty recollections of history. The Hudson Bay route to Lake Winnipeg and interior North America, shorter and cheaper, was now the obvious choice, and York Factory on Hudson Bay replaced Fort William in function and importance.

The accomplishments of the Nor'Westers, however, had not been limited to Lake Superior. Some had gone on westward to leave their personal marks in North American history.

In 1789, Alexander Mackenzie canoed the great river that was to bear his name, from Great Slave Lake to the Beaufort Sea and Arctic Ocean. Simon Fraser in 1808 fought through defiles in the Rocky Mountains down the river he mistakenly thought was the Columbia, a river eventually named the Fraser, to the Pacific. David Thompson coursed down the real Columbia in 1811 to provide a remarkably accurate survey of the great river, establishing the first navigable Northwest Passage to the Pacific and his fame as a charter of rivers.

All these men, and many more of the North West Company,

blazed early trails across the continent, always through the gateway of the North Shore of Lake Superior.

Not all activity on Lake Superior ended with the demise of the company. But henceforth, business was conducted on a shorter and more direct route, from James Bay south to the North Shore.

Now in the ownership of the Hudson's Bay Company, Fort William continued to function as a supply depot, stopover point for important travelers and company officials, and an early commercial fishing establishment. The Gladman-Hind expedition of 1857 passed through Fort William on its way to the Red River, seeking an efficient passage to the west—not for furs now, but for immigrants to settle the western prairies.

With the increased use of larger sailing schooners on Lake Superior, Fort William suffered because of its location by the shallow waters of the lower Kaministiquia River, which could not accommodate the larger ships. But a few miles to the north, a rocky ridge ran down from inland to jut out into Thunder Bay and provide a deepwater dock for ships. This location, competing with Fort William, became "The Landing," then "Prince Arthur's Landing," and later the city of Port Arthur.

It was the coming of the railroads and the completion of the Canadian Pacific Railway in 1885 that administered the *coup de grâce* to Fort William. The site of the old post was taken over by the Canadian Pacific.

Eventually engulfed, Fort William's identity was lost by the developing city of Fort William, and the great emporium on the banks of the Kaministiquia River fell into a forgotten realm.

Trading in furs at the western end of the North Shore was notable by reason of a most significant connection: Lake Superior to the Mississippi Valley. In a day when almost all traffic in the Northwest was by birchbark canoe, this mostly water connection made possible a vast increase in new territory open to the fur trade.

Although Sieur du Lhut in 1679 pioneered this route for later travelers, the French were never significantly involved here in fur trading. Joseph, son of explorer La Vérendrye, in the 1750s operated a post on the south shore and probably used the

St. Louis River route to the Mississippi, but there is no evidence of French posts in the St. Louis River area.

Following the French and Indian War, however, British trading activity came quickly to the head of the lake. Free-trader Alexander Henry established a post in 1765 at the La Vérendrye site; in 1784, a North West Company post was established at the mouth of the St. Louis, the first on the river; and in 1793, Fort St. Louis was erected by North West Company partners on the great harbor behind Minnesota Point, in what is now Superior, Wisconsin. Although the time followed the American Revolutionary War, the British North West Company operated its posts on American soil for several years without interference. Its trading activity based at Fort St. Louis flourished for a period of more than two decades, using St. Louis River routes to posts and wintering houses in the hinterlands.

One route led upstream on the St. Louis across the Lake Superior-Hudson Bay divide to connect with the old Pigeon River route. Another route, the most significant one, led to the vast upper Mississippi Valley.

Both routes at first followed up a torrential gorge in the lower St. Louis, totally impassable to any river traffic. This reach thus necessitated a rough detour by land—also called Grand Portage. The trail, seven miles long, included a single steep climb of more than 100 feet and required three days of effort. Another portage followed; its length, one and a half miles. Farther upstream, at what is now the town of Floodwood, Minnesota, the two routes separated—the northern route continuing on the St. Louis toward connection with the Pigeon River track, the Mississippi route branching to the west.

Owing to unique geologic development during times of glacial retreat, the swampy headwaters of two small streams formed a natural divide between the Lake Superior and Mississippi drainages. One—the East Savanna River—flowed east to the St. Louis; the other—the West Savanna River—flowed west from the divide toward the Mississippi. Today this portage is the focal feature of Minnesota's Savanna Portage State Park.

Traders of the North West Company labored over the St. Louis and Savanna portages for more than twenty years. Large quantities of beaver pelts were removed from the Rainy Lake area and the upper Mississippi through Fort St. Louis.

But eventually the advantage enjoyed by the Nor'Westers came to an end—through the influence and effort of American John Jacob Astor. He was presently to garner all fur trading on the American North Shore, and in much of the rest of the United States as well.

Astor was head of the American Fur Company, chartered in 1808. In 1811, he joined with several Montreal merchants in the North West Company to form the South West Company. By this time, however, the United States had laid firmer claim to its North Shore property, and British interests were disallowed. But since Astor himself was an American citizen, South West Company operations of the old North West posts on American land were tolerated. Later, Astor bought up his British partners' interests and assimilated them into his American Fur Company, a masterly stroke.

Now Fort St. Louis was deemed insufficient for expanded purposes of the American Fur Company. It was abandoned in favor of a new post, Fond du Lac, constructed in 1817 a little upstream on the St. Louis at the head of navigation.

The Fond du Lac post never carried the military title of "fort," but it became an expansive settlement, operating for twenty-five years as the headquarters of the American Fur Company's *Fond du Lac Outfit*, serving many inland trading and wintering posts. In 1832, Fond du Lac became headquarters and packing plant for the American Fur Company's commercial fisheries enterprise on the North Shore.

In the same year, Henry R. Schoolcraft stopped over on his way to christen Lake Itasca, headwater of the Mississippi.

But the American Fur Company's part in the North Shore fur trade was not to last much longer. A great economic depression in 1837 was disastrous to the company; its commercial fishery collapsed in 1842, and so did the company itself.

The same year, the Fond du Lac post was purchased by the Missouri Fur Company, which continued some trading activity at Fond du Lac until 1848.

The mouth of the Michipicoten River had long been a center for communication and travel on Lake Superior. Many archaeological diggings at the site indicate important use by prehistoric peoples. Early Americans probably used the inland route up the river and over the height of land to James Bay.

The French first established a post here in 1725. As the inland route had served the English post, Moose Fort, on James Bay, the post at Michipicoten served the French and British in turn, largely to interdict Indians trading with the Hudson's Bay Company. Michipicoten was to serve in that capacity for nearly a hundred years.

This first French station was established within their *postes du nord* policy—a matter of building posts at the mouths of North Shore tributaries for the purpose of capturing Indian trade to James Bay. Included were posts at Pic River and Nipigon River, from both of which routes led northward. For thirty-five years the French and English competed for the natives' furs in the Michipicoten area.

At the close of the French regime in 1763, French-English competition on the Michipicoten route ended, too. But then the Michipicoten continued as the scene of competition between the two British companies, North West and Hudson's Bay.

At the mouth of the Michipicoten River, wide, flat, sandy land facilitated greatly the establishment of trading posts, later missions and settlements. Upstream, the inland route followed up the Michipicoten, after passing initial falls and rapids, to where the river was largely flat water, and on to its headwater Metagomy Lake. Then, across the height of land portage to the Hudson Bay drainage, the route passed into the valley of the Missinaibi and Moose rivers, thus to James Bay and the Hudson's Bay Company post of Moose Fort.

Altogether, it was considered a difficult route, with rough rapids and many falls, of some 325 miles. In the historic period of its use, this water highway served explorers and fur traders as the principal connection between Lake Superior and Hudson Bay, for a century and a half. Early Americans had undoubtedly used it for millennia.

After the formation of the North West Company, both Hudson's Bay Company and the Nor'Westers used the route intensively. Both built additional posts, moving frequently to match the competitor's locations.

Actually, by locating and trading north of the height of land that separated Lake Superior and Hudson Bay drainages, the

North West Company was trespassing, acccording to long-established rights of the Hudson's Bay Company. But, since the Michipicoten route was obviously Hudson's Bay's main connection to Lake Superior, the Nor'Westers were willing to take chances in trying to foil their rivals. It appears that in the wilderness, far from the halls of authority in London, the more immediate forces of ambition and survival mattered most.

In this rivalry the Hudson's Bay Company gradually became dominant. The violent disputes became a factor in the amalgamation of the two companies in 1821, and after that, Hudson's Bay Company hegemony was complete.

The company then established its *Lake Superior District*, within which were all the old trading posts along the Canadian North Shore, with headquarters at Fort William. Michipicoten assumed even greater importance in 1827 when district headquarters was moved there from Fort William. Michipicoten House now became the administrative center of all Canadian fur trade on Lake Superior.

The route of travel from Europe to the Superior country was now shorter and cheaper—via Hudson Bay. Michipicoten's central role in the trade was to extend for twice as long as Fort William's.

For the next half-century, furs were collected at Michipicoten from around the Canadian North Shore, transferred to the inland brigades, and sent north in June of each year to James Bay. It was a prosperous time. For with amalgamation, peace had replaced the intense rivalry between Lake Superior and Hudson Bay that had existed—between French and English, later between the Nor'Westers and the Honourable Company—since the days of Du Lhut's first forts.

More than fur now commanded the attention of the new occupants of Michipicoten. Missions treated spiritual needs of the natives. Agricultural endeavors assisted the provisioning of food; the growing of potatoes and such livestock as cattle, sheep, and pigs brought an almost luxurious cuisine.

Fresh fish, too, contributed a welcome addition to dinner tables. Whitefish and lake trout were taken in abundance at the mouth of the river and at more distant points: Gros Cap near Sault Ste. Marie, Gargantua Bay, offshore Michipicoten Island. Fishing was expanded to a commercial enterprise, with trout and whitefish marketed through the Sault to points down the lakes, as well as supplying other Hudson's Bay Company posts along the shore.

Other endeavors of civilization were introduced at Michipicoten during these final days of the Lake Superior fur trade. The settlement became a major construction center for keel boats, sturdier and larger, used on the route to James Bay. It also became a center for manufacture of other kinds of lake and river craft, and repair facilities were kept busy working on the company's fleet of Lake Superior schooners. Local artisans filled other essential manufacturing roles—carpenters, coopers, black-smiths. The large tin works fashioned kettles, plates, teapots, and lamps for many Hudson's Bay Company posts; Michipicoten's tin products were known widely for their high quality.

Later decades brought two major events that changed the role of the Michipicoten establishment and eventually brought about its decline. In 1855 locks at the St. Marys River rapids allowed passage for schooners, later steamers, from the lower lakes. And in 1885, the Canadian Pacific Railway was completed, crossing the Moose-Missinaibi route, and resulted in the establishment of settlements and towns on the old canoe trail.

Hudson's Bay Company ceded its granted authority and its lands to the Canadian government in 1869. Michipicoten continued to serve as the company's district headquarters, however, until this function was transferred to Red River in 1887. And in 1897 the Province of Ontario established its headquarters for the Michipicoten Mining Division at the old post, presaging the next stanza in resource exploitation around Superior's northern coast. Michipicoten House, the last fur trade headquarters on the North Shore, finally closed in 1904.


The decline of the fur trade on the North Shore occurred when it did—during the mid-1800s—for several reasons. The depression of 1837 was one, reducing trade across the continent. European fashion about this time shifted from hats made of beaver fur felt to less expensive silk; the British army abandoned its traditonal beaver hats; furs of the South American nutria, similar to beaver, were imported as a substitute; locally, the fur of the muskrat replaced that of the beaver. Above all,

79

however, was the decline in beaver populations, resulting from a century of intensive trapping to a conditon of near extirpation that would require many decades for full recovery.

And the great posts—what happened to them in ensuing decades?

Michipicoten became, first, a mining and shipping center and later, with attendant town of Wawa, a focus for recreation and tourism for the eastern shore. At the other end of the lake, Fond du Lac served as a nucleus for Minnesota's first permanent settlement—the town of Fond du Lac—and later the city of Duluth.

The elegance of the Kaministiquia River entrepôt was recaptured in 1975 in the reconstruction of Old Fort William, now on a point upstream where in 1816 Lord Selkirk's armed soldiers had made their camp. Operated by Ontario's Ministry of Tourism and Recreation, Old Fort William each summer reenacts for modern explorers the bustle of the old rendezvous, on the banks of the river that once carried the old brigades.

And in 1960 the site of the Pigeon River's Great Carrying Place was established as Grand Portage National Monument, in the United States National Park Service. The reconstruction presents the depot as it was in the heyday of the North West Company's trade, 1797. Three flags in order—French, English, American—now flutter against the background of surrounding hills, above a stony beach upon which more than 250 years ago Sieur de la Vérendrye first set foot, to open the North Shore's era of fur empires.

In the middle of the fur trade era—roughly 1700 to 1850—occurred the American revolution, earning independence for the new United States of America. Now from coast to coast would run an international boundary—directly through the center of the fur trade region. Setting the border's location was at first a matter of great contention between Great Britain and the United States, and fur trade matters were to play a significant role in this dispute.

It was a formidable job, for most of the area through which the boundary was to run remained an unsettled wilderness, of which the lack of geographic knowledge was outstanding. However, the treaty negotiators had a supposed great advantage: an amazing chart which came to be known as "Mitchell's Map."

Native-born in 1711 in the colony of Virginia, residing in England for his health, John Mitchell had been educated as a physician. But he was also an amateur botanist, zoologist, and cartographer. In fact, it appears that making maps was his all-consuming hobby. The map he drew of North America and published in 1755 was remarkable for its detail—accurate or not—and it was relied upon heavily by the treaty makers in 1783.

Now diplomats in setting borders prefer the use of rivers and lakes, easily defined and permanently located. So, naturally, when the negotiators considered the new border they searched for lines along known routes of water travel. Obviously, the boundary was to come down to the North Shore of Lake Superior somewhere, and the voyageurs' route from Lake of the Woods to Lake Superior seemed a most natural water track. Such a route was clearly indicated on Mitchell's Map, and a bold red line was drawn across the map accordingly.

That track was further described in the treaty document: "through Lake Superior . . . to Long Lake; thence through the middle of said Long Lake, and the water communication between it and the Lake of the Woods." No one really knew where "Long Lake" was, except that it did appear on Mitchell's Map, on the North Shore. And in 1783 the red line went right through it.

In 1822, a first international commission proceeded to settle the boundary's location, but disagreements soon arose. A major point of contention was the location of "Long Lake." On the voyageurs' line of "water communication," the mystical lake could conceivably have been one of three possibilities: the estuary of the St. Louis, Pigeon, or Kaministiquia rivers. If it were the southernmost, the St. Louis, the British would gain the most; if it were the Kaministiquia, the northernmost, the Americans would gain the most. The Pigeon seemed a logical compromise; and most of the persons involved agreed that the Pigeon River route was intended. Surely, Mitchell's Map, with its bold red line, would settle the matter. But by this time, thirty-nine years after the treaty, the original map was lost, never to be found again. Disagreements went from bad to worse. Separate commissioners' reports were submitted and filed in obscurity.

The boundary had to be settled eventually. Another twenty

years later, negotiations were opened afresh. Daniel Webster, secretary of state for the United States, and Lord Ashburton, special envoy for Great Britain, in 1842 played the central roles. The two statesmen, old acquaintances, quickly came to an agreement on the choice of the Pigeon River route as the intended border, as well as on other points. The Webster-Ashburton Treaty, establishing the boundary between Canada and the United States from the Atlantic to Lake of the Woods, was ratified by both countries in late 1842.

In the negotiations that finally settled the long-disputed boundary, resolution of the "Long Lake" question came easily. Nevertheless, Webster cited the land between the St. Louis and Pigeon rivers as "disputed," a land of great mineral value, and thus claimed his effort to be a huge diplomatic success.

It is true that during the period of negotiations, valuable mineral deposits were found elsewhere in wild lands of the Lake Superior region. Ergo, the great area "won" by Webster's treaty-making efforts must also contain valuable minerals—but no one knew just exactly *which* minerals, not the least of whom, apparently, Daniel Webster. Certainly, no one thought of a mineral as unromantic as *iron*—or that the name *Mesabi* would ever bring to mind anything but an imaginary giant in Ojibway legendry.

# Iron Mountains and Other Lodes

The sediments laid down two billion years ago in the Superior region, beneath the Animikie sea, now in mid-nineteenth century lay only a short distance below the feet of the new occupants of the North Shore. Some of these sediments, hardened to rock, contained about 30 percent iron, and their extent was vast. They were later to take a name that reflected similarly appearing rock in the Taconic Mountains of Massachusetts—taconite.

The iron content of taconite was too low to be used in iron-smelting blast furnaces. But in limited bands, inland from the northwestern coast of Lake Superior, a natural concentrating effect had occurred possibly a little more than 100 million years ago. Owing to oxidation and leaching by surface waters, the iron content of the taconite was increased nearly threefold. Taconite was the "mother ore"; the enriched daughter materials were to be called "natural ores."

These natural ores lay along the slope of an ancient granite mountain, the old roots of which lie today in the form of a massive ridge, fifty miles long and up to 400 feet high in places, running roughly in an east-west direction. This range of hills was such an impressive aspect of the local terrain that it held a re-

*Wawa Lake*

vered place in the legendry of prehistoric Ojibway: the formation was the home of a great sleeping giant.

And so the old mountain came to be called Giants Range. But the Indians kept their older name for the giant that slept in the great range: *Mesabi*.

Iron, of course, was only one of several valuable minerals that lay buried in the ancient Precambrian rock rimming the Lake Superior basin. By the time Mesabi had fully yielded its richest treasure in iron, the rush for other metals—copper, silver, gold— had embraced a full century.

Although the locations of mineral wealth of the North Shore were unknown to the white newcomers in the

*Wawa Lake*

region, such wealth was by no means unsuspected.

Had not Daniel Webster, in securing for the United States that part of the North Shore between Duluth and the Pigeon River, in the boundary-setting treaty of 1842, made the titillating suggestion of great mineral resources? Why not around the entire North Shore?

A literal fury of prospecting, land speculation, and transportation development lay in waiting.

In anticipation of a mining boom, Canada instituted a system of prospecting "locations" along the shore. Each location was two miles along the shore, five miles deep. In 1845, 160 applications were received by the government, covering much of the Canadian shore; in the next year, twenty-seven locations were granted.

The prospector waited for the removal of one further obstacle: ownership of the land by the Ojibway, which prevented unrestricted mining. This final obstacle was eliminated by two treaties of land cession—Canada's Robinson-Superior Treaty in 1850 and the United States Treaty of 1854—which, with the exception of a few reserves, opened the North Shore regions of both countries to mineral prospecting. In fact, both treaties were stimulated largely by the white man's pecuniary interest in hidden minerals.

Copper was the first valuable mineral to be known in quantity around Lake Superior. Indeed, 5,000 years ago prehistoric peoples had mined copper from Michipicoten Island and Isle Royale; early Jesuit missionaries had reported the spiritual relationship between Indians and copper. Alexander Henry had explored known copper deposits on Michipicoten Island and Mamainse Point in the 1770s. By the 1840s, copper on Keweenaw Peninsula on the south shore—the greatest deposit of native copper in the world—was being exploited. The boom in copper prospecting charged the Lake Superior region with mining fever.

In addition to copper mining activities on the Keweenaw Peninsula and Isle Royale, many attempts at copper mining were made along the Canadian shore—at St. Ignace Island, Michipicoten Island, Little Pic River—but eventually with negligible results. During a few years following 1879, a mining village developed at Mamainse Point, involving several mining companies, but all were closed five years later. Michipicoten Island, which had seen intermittent copper mining activites since the 1860s, continued to yield modestly into the mid-1900s, with continuing digging of experimental shafts, sampling of ore, and mapping of possible copper-bearing formations. Copper extraction had continued sporadically on Isle Royale into the latter part of the 1800s.

But around the North Shore generally, no copper deposits considered large by modern comparisons were found, and no fortunes were made. In effect, by the time the two land cession treaties were completed in midcentury, the copper boom had nearly fizzled out.

Spurred by events elsewhere, prospecting continued. Locks at Sault Ste. Marie had opened in 1855, greatly improving transportation of heavy loads by water; steel tracks were being pushed over other wilderness regions of the continent. Most important, gold and silver finds in the mountains of the West were making fortunes, spurring the search for the noble metals elsewhere.

The lure of bright metal infected prospectors on the North Shore as anywhere else. The Montreal Mining Company bought up locations to accumulate huge holdings. And lonely men with pick and shovel continued to comb the ancient Precambrian rock—this time to search for the brightest metals of all.

The Ojibway in the Thunder Bay region had their silver legends, and to the prospectors who had patented two-by-five-mile locations along the North Shore, the presence of silver was not unexpected. As early as 1847, or possibly 1846, small amounts of silver and copper were discovered in a vein at Prince Bay, between Fort William and the Pigeon River, leading to Canada's first mining venture along the North Shore.

The real rush for silver began with the discoveries in 1866 on the diabase-capped "mountains" of the Nor'Westers Range west of Thunder Bay. These finds led to the establishment of the Rabbit Mountain Mine, and soon after, the Beaver, Badger, and Porcupine mines, then Shuniah and Silver Mountain. Silver mining flourished, yielding close to two million dollars worth of the metal. But toward the close of the century, silver prices fell, the richest veins in the old diabase sills ran out, and mining ceased.

These endeavors on the Nor'Westers Range were greatly overshadowed by the silver lodes discovered in a tiny, unnamed rock three-quarters of a mile off the shore of Sibley Peninsula. Measuring a scant ninety feet by ninety feet, rising not ten feet above Lake Superior's surface, the rock was soon to be given the appellation *Silver Islet*, so rich were its veins. For fifteen years, it was to hold the richest-producing silver mine in the world.

Prospector Joseph Woods in 1845 patented a location that included the little island, and he sold it at a modest profit to the Montreal Mining Company the next year. The company, still mainly searching for copper, accumulated many mining locations along the North Shore, but this one, known officially as "Wood's Location" in company records, was to become eminently famous twenty years later.

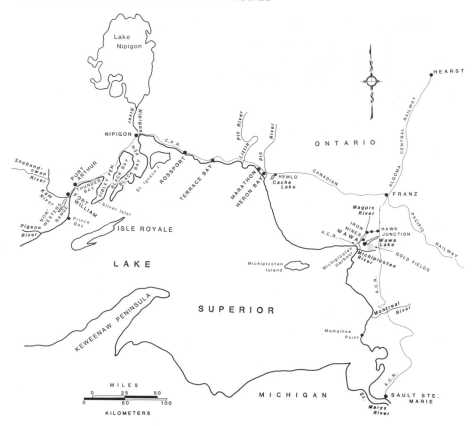

*Major mining locations on the northern and eastern portions of the shore. These included silver mining in the Thunder Bay area, gold mining and later iron mining in the Wawa area, and the current gold operations near Hemlo.*

While camped along the shores of the peninsula in July of 1868, members of a Montreal Mining survey party landed on the tiny rock intending to use it as a transit base. The shape of the little island resembled, they thought, a human skull, and on their charts the surveyors labeled it "Skull Rock." But instead of a transit base, the party found veins of native silver. After an exploratory blast with black powder, silver was plainly visible in a vein running out beneath the clear waters of Lake Superior. In one afternoon, more than $6,000 worth of native silver specimens were simply pried and hacked out of the rock. "Skull Rock" was erased from the surveyor's chart and replaced by "Silver Islet."

88

Silver Islet was formed of slates that resulted from silt deposits beneath the Animikie sea; it is not the type of rock to contain rare metals or ores. But long after the Animikie sediments had settled and hardened, the Keweenawan volcanism drove up tendrils of magma, to solidify as dikes in the overlying slate. These diabase dikes contained native silver, as well as veins of quartz and calcite containing silver ore. A fault occurred across the tiny island, displacing the dike and exposing the bright metal—a fault that appeared to extend northwestward into Sibley Peninsula, raising the speculation of undiscovered silver lodes on the mainland.

In this summer of discovery, the Montreal Mining surveyors extracted only their specimens for assessment. Serious mining began in the next year, with the arrival of heavy equipment and machinery. But 1869, as it turned out, was a year of particularly violent activity on Lake Superior, and the islet—at times completely inundated by storm-driven waves—took a beating. The storm damage presaged even greater troubles ahead. Yet the following winter brought respite, and mining continued.

Again, in the next year, Lake Superior storms plagued mining operations—to the point where the company sold out its holdings. Bought and operated by an American syndicate, the mining enterprise continued, now under the name "Silver Islet Consolidated Mining and Lands Company."

Heading the company of new owners was Major Alexander Sibley, brother of Minnesota's governor and namesake-to-be of the peninsula's geological formations. But it was a young copper mining engineer, William B. Frue, who as mine superintendent fought Lake Superior's worst tantrums to a standstill. Time after time, storms battered the islet and wrecked breakwaters, cofferdam, and mine machinery. A monstrous storm in November 1873 in addition took two lives. Time after time, Frue repaired the damage and made stronger fortifications. Battling the elements, repairing damaged buildings, and replacing destroyed equipment, Frue and his crews continued to extract silver. Also required was the constant pumping of water from underground shafts that went as deep as 1,230 feet below Lake Superior's surface.

The Silver Islet mine yielded its fortunes for fifteen years, during which time it was billed worldwide as the richest mine on earth and "an island of silver." Its ore was assayed as the richest in silver ever found.

In 1870, families arived to create the settlement of Silver Islet Landing on the peninsula mainland. New buildings were erected: homes, school, churches, offices, jail. The "Landing" developed a community that was to serve the Silver Islet operations for fourteen more years, until the mine closed.

In the fall of 1884, a shipment of coal from across the lake that was to fuel the pumps failed to arrive—frozen in for the winter. The pumps stopped, the shaft was flooded, and the mine closed. The battle, pitting a tiny rock of silver against the elements of Lake Superior, was over.

At the end, the value of silver extracted had reached three and a quarter million dollars, creating fortunes in its day, but almost totally balanced by the extreme costs of maintaining the tiny island fortress against the storms of northern Lake Superior.

Silver Islet produced more than all the other Thunder Bay mines in combination. Some of these others continued to produce for a decade or more, but by the early 1900s the silver boom on the North Shore was over.

Silver Islet Landing, on Sibley Peninsula opposite the islet, had served as the mine's headquarters and support center during the silver heyday. The community became a summer resort colony, remaining today as a private enclave near the end of the peninsula's Sibley Provincial Park.

The first gold rush in the country north of Lake Superior ironically lacked, for all practical purposes, one of the more important elements: gold.

In the mid-1800s, much interest in iron ore was evinced around the lake. The Minnesota legislature ordered a survey of the western North Shore region, particularly for iron, and in 1865 the governor appointed state geologist Henry H. Eames as director of the survey. And although Eames did in fact report large exposures of iron ore—in the Vermilion Lake area—he apparently was more interested in the brighter metals. When he reported gold-and silver-bearing quartz in the vicinity of the lake, and submitted specimens for assay with favorable reports, the Vermilion Lake gold rush was on.

Immediately, prospectors scrambled over the area in a wild stampede, and Vermilion Lake became the focus of gold fever. Cabins and houses were quickly erected, and the town of Winston City appeared, complete with general store, sawmill, and fourteen homes. During that first year and the next, at least fifteen mining companies were established in the Vermilion area, financed from St. Paul and eastern cities.

To provide ready communications and access for heavy mining machinery, the Vermilion Trail, eighty-four miles long, was blazed through the wilderness from Duluth to the Vermilion "gold fields." With the ground frozen in winter, teams struggled up the trail with heavily loaded wagons, including parts for four stamp mills.

Hopes ran high. But returns in the form of gold dust did not, and by the end of 1866 most miners and prospectors had abandoned their claims and cabins. Only a few of the larger companies remained to continued their attempts, unsuccessfully, the next year. Gold was, in fact, present but in quantities much too low to be profitably extracted. Winston City disappeared.

And what of Eames's samples? "Salted" gold—deliberately placed in the area to give the appearance of rich deposits—was a possibility, but not probable, for no one stood to profit. Most likely, Eames, with his mind on gold, unconsciously picked out rich samples that were not representative. He never found his pot of gold.

With the rush over, the shores of Vermilion Lake were quiet once more. And, although the iron ore that Eames had sought along those shores had not gone unnoticed, even by others, the lake shores were to remain quiet for another twenty years, awaiting the development of the Vermilion Iron Range.

Real gold—in commercially significant quantities—was first discovered in the North Shore country in the winter of 1870-71 by two Hudson's Bay Company Indian trappers near Shebandowan River, a tributary of the Kaministiquia, seventy-five miles west of Fort William. Samples were brought in by the trappers to the Hudson's Bay post at Fort William, and assays showed both gold and silver. So a rush was on to the Shebandowan country. At several locations in the area and under a succession of owners and mining companies, the Shebandowan gold field produced until 1936, when operations closed.

The greatest-paying gold rush on the North Shore, except for modern finds, was that in the Wawa Lake area, beginning in 1897. In the spring of that year, an Indian couple—William Teddy and his wife—stopped for lunch along the shores of Wawa Lake and found chunks of gold-bearing quartz. The find set off another stampede of thousands of propectors who overran the hills and forests and the land along all rivers and streams of the region. In the fall of the same year, Ontario's Bureau of Mines at the old Michipicoten fur post of the Hudson's Bay Company set up its first mining division office. The office soon recorded more than 1,700 mining claims.

The town that was later to take the name Wawa developed as a gold boom town, rapidly completed with boarding house, stores, pool room, bakery—and the "Balmoral Hotel" boasting a forty-foot-long mahogany bar. At Michipicoten Harbour, on Lake Superior by wagon road seven miles west of Wawa, the old settlement of fur post and mission boomed again, now with increased ship traffic, including hopeful prospectors.

The mines that were developed during this early boom were mainly located in the hilly, forested area between Wawa Lake and the Michipicoten River to the south. Some of the more important of these, opened in the closing years of the century, were the Grace Mine (the largest producer), Sunrise, Minto, Jubilee Lake, and Mariposa. Most of these early mines operated for two or three years, after which the easy gold was worked out. Speculation was rife, and land deals—both legitimate and bogus—made the land and claim brokerage business more profitable than gold mining. But several million dollars in gold were extracted.

By the first few years of the new century, this boom was over, and in 1911 the settlements were ghost towns.

With improved technology, several mines were reopened in the 1920s and 1930s. The Grace Mine was reopened in 1926 and renamed the Darwin; the Sunrise, Minto, and Jubilee followed in the next five years. The Mariposa reopened as the Parkhill, to gather around it the largest of the new settlements. This second rush was profitable, too, so that while much of the rest of the world suffered economically in the 1930s depression, the Wawa area flourished. By the end of the decade, the Wawa gold fields had produced another three and one third million dollars.

In the year of the first big gold rush, 1898, prospector Ben

Boyer had been searching for pay dirt in the hills north of Wawa Lake. Boyer and his partner, Jim Sayers, found what they thought were specimens of gold, but which turned out to be merely fool's gold—iron pyrite. In the fall of the same year, Alois and George Goetz, two brothers from Sault Ste. Marie, Michigan, found in the same area an outcrop of "brown ore," rich in iron.

These finds triggered the richest metal mineral exploration along the Ontario North Shore so far. Not gold in these hills now, but a virtual mountain of iron.

The iron discoveries spurred the dreams of entrepreneur Francis H. Clergue of Sault Ste. Marie, who had great plans for industry on the eastern shore. To water power at the Sault and pulp forests to the north, now could be added yet another vital element: iron for the production of steel. Clergue immediately bought up Boyer's and the Goetzes' rights. The addition of iron in the closing years of the nineteenth century made possible the full development of Clergue's huge industrial complex at Sault Ste. Marie, to be known familiarly as the "Clergue Syndicate."

The Helen Mine, named after one of Clergue's sisters, established in 1898 and soon to become Canada's largest iron ore producer, was the first substantial iron mine in the country. The lake along which the finding was made was soon named Boyer Lake, the massive hill of ore, Helen Mountain.

One of the first problems faced in development of the mine was transportation. And so Clergue's Algoma Central Railway was incorporated in 1899, and in the same year construction began from two points: first, from the Sault northward and, second, from Michipicoten Harbour east to Helen Mountain. The shipping facility of Michipicoten Harbour was constructed, and, with the purchase of four steamships in 1900, the Algoma Central Steamships line was formed, to carry ore from the Helen Mine.

During 1899-1900 the rail line from Michipicoten to the Helen Mine was blasted through the hard Precambrian Shield. With supplies and men shipped by steamer from the Sault, the rail line by time of completion comprised five locomotives, 100 ore cars, and twenty miles of rails. Michipicoten Harbour included a huge ore dock, commercial dock, warehouses, offices.

Algoma Central steamers soon began carrying 1,000 tons of ore per day, during the ice-free navigation season, to Clergue's new steel plants at the Sault. By 1910, rails from the Sault had reached Hawk Junction, 138 miles north from the Sault, and a connection was completed by 1912 from the Helen eastward. Thereafter ore could be transported from the mines to the Sault year-round.

The Helen's ore was high-grade. Mining was by both open pit and underground tunneling, and 400 miners labored at extracting the ore. A supporting settlement developed, complete with general store, post office, public school. Cattle shipped by steamer from the Sault and rail from Michipicoten Harbour provided fresh meat and dairy products; the free-ranging cows often wandered into the streets of town.

The Helen Mine produced essential quantities of iron for the First World War, but afterward it was unable to compete with the iron mines of Minnesota. The Helen closed in 1918, when the highest grades of ore ran out; the ghost town at the mine site was destroyed by fire in 1921.

In addition to the Helen, other mines soon opened in the same area. The Josephine, discovered by Alois Goetz ten miles northeast of the Helen and at about the same time, also contained high-grade ore and was mined by underground methods. The Magpie Mine was developed twelve miles north of the Helen, requiring another rail spur that was completed in 1911, first across the Magpie River by a huge wooden trestle. The Magpie's ore graded too low for use in blast furnaces, and a roasting plant that beneficiated the ore to sinter was constructed at the Magpie River crossing, Canada's first sintering plant, operating until 1921.

With the closing of both the Helen and the Magpie, iron mining in Canada came temporarily to a halt. But by the early 1920s, mines in the Wawa area had produced about four million tons of iron ore. Between 1924 and 1939, the Ontario mines produced no iron ore at all.

Stimulated by demands of another impending world war, and as well by Canada's "Helen Mine Iron Ore Bounty Bill" that subsidized mining by two cents per ton of ore, the New Helen mine was opened in 1939 near the old, by Algoma Steel, extracting the more abundant but lower-grade ore. Part of the New Helen's ore body lay beneath a lake, which was pumped out to permit mining operations. A sintering plant was built at the site of Wawa, and by the mid-1950s, the New Helen was producing one and a half million tons of sinter per year.

The town of Wawa received the blessing of its official name in 1960, along with the completion of the Trans-Canada Highway nearby. Built on gold and iron, Wawa flourishes now as a major tourism and recreation center, revealing little of the feverish activity of prospecting, land speculation, and rumbling ore cars that once held sway in Canada's first large mineral enterprises.

By the time the major iron mining in the Wawa area tapered off, soon after the First World War, iron ore extraction from mines in Minnesota had far overshadowed ore production on the eastern shore. The Vermilion and Mesabi ranges, particularly the latter, had become the largest and richest source of iron in the world.

The many prospectors involved in the Vermilion "gold rush" of 1865 included some who were aware of the potential of iron. One of these, George Stuntz, sighted the ore at Vermilion Lake and promoted its future value; it was Stuntz who, in the late 1860s, had directed the construction of the Vermilion Trail. Financier Charlemagne Tower formed the Minnesota Iron Company in 1882, authorized the construction of the sixty-eight-mile-long Duluth and Iron Range Rail Road from Two Harbors on Lake Superior to Vermilion Lake (two years later the line was extended to Duluth), and built the great ore docks on Agate Bay (one of the "Two" Harbors).

And on an early summer day in 1884, the first trainload of Minnesota iron ore left the Soudan Mine, destined to be one of Minnesota's most famous mines. The first shipment from the Soudan, in ten wooden cars pulled by a woodburning engine, was 240 tons. Later that year, two steamers together carried nearly 3,000 tons to blast furnaces on the lower lakes—the first steamship ore loads to leave the North Shore.

This 3,000-ton load, so puny in retrospect, presaged incredible shipments of ore in the next half-century, from the Minnesota mines.

The Soudan operations were by open-pit mining at first, but the hard Vermilion ore was deeply located, and mining was car-

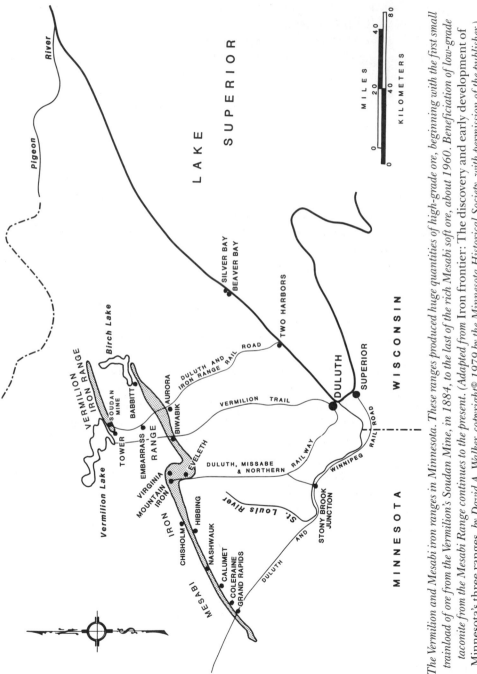

The Vermilion and Mesabi iron ranges in Minnesota. These ranges produced huge quantities of high-grade ore, beginning with the first small trainload of ore from the Vermilion's Soudan Mine, in 1884, to the last of the rich Mesabi soft ore, about 1960. Beneficiation of low-grade taconite from the Mesabi Range continues to the present. (Adapted from Iron frontier: The discovery and early development of Minnesota's three ranges, by David A. Walker, copyright© 1979 by the Minnesota Historical Society, with permission of the publisher.)

ried out underground from shafts that eventually reached over 2,500 feet deep.

In early years, mules were used underground to haul ore though the horizontal drifts to hoists that then lifted the ore to the surface. Kept underground in subterranean darkness for long periods, the animals were brought up at six-month intervals to slowly reacclimate them to surface light.

In the first year of its operation, the Soudan Mine produced over 60,000 tons; in its second full season, 1886, nearly a third-million tons were shipped. By the time the Soudan Mine closed in 1962, it had produced sixteen million tons of high-grade ore. At the time of its closing, the owners, United States Steel Corporation, donated the mine and surrounding lands to the state of Minnesota (actually "sold" it for $1.00) for a state park—today's Tower Soudan State Park, a historical rather than recreational park. Visitors are lowered down a 2,400-foot-deep shaft and then given a train ride along a horizontal drift to the stope where ore was mined.

Vermilion Range ore was hard rock—high-grade ore. The range extended roughly eastward from Vermilion Lake to Ely, eighteen miles. In total, eleven mines on the Vermilion Range, including the Soudan, produced 100 million tons of ore, from the first tiny shipment from the Soudan in 1884 to the closing of the last Vermilion mine in 1967.

But even this immense quantity was eclipsed by the ore production of the Mesabi Range, which was more than twentyfold greater than the Vermilion's.

When Henry Eames landed on the shore at Beaver Bay in September of 1864 on his survey for iron, he engaged Christian Wieland, engineer and surveyor of Beaver Bay, as a guide to lead the way to Vermilion Lake. Inland a few days later, while crossing the forested uplands near Birch Lake, Wieland's compass suddenly went astray. "We're standing on iron!" he exclaimed. "To hell with iron!" Eames replied. "It's gold we're after!" So close did the state geologist come to one of the richest deposits of iron on the North American continent.

Wieland, however, kept the discovery in mind, and he returned to the site of his magnetic find and brought out samples

97

of high-grade iron ore. With iron mining friends from the south, Wieland formed the Ontonagon Syndicate, which selected Peter Mitchell, one of their members and an experienced prospector, to lead a party to explore Wieland's discoveries. In 1870, Mitchell's party sank many test pits in the high Giants Range country south of Birch Lake; they reported back with more high-grade samples and a tale of Mitchell's "iron mountain."

In 1874, the Ontonagon Syndicate incorporated the Duluth and Iron Range Rail Road Company but did not build it. (Later, Charlemagne Tower acquired it and ran the line to Two Harbors to serve his Soudan Mine on the Vermilion.) In 1882, the syndicate formed the Mesaba Iron Company, but it did no mining. Instead the group began to sell its land holdings, albeit at a good profit, and not long afterward the syndicate dissolved. With Mitchell's death in 1908, the last member was gone.

What had happened?

The syndicate had been on the edge of a great discovery—actually, right off the eastern tip. Theirs was the first serious exploration of the great Mesabi—close, but a clear miss. What they had found instead was an enormous "mountain" of taconite rock, worthless at the time. Mitchell's high-grade samples apparently had been handpicked. It was to be twenty-five years after Christian Wieland's discovery in 1865 that the big strike was made.

Meanwhile, Lewis Merritt of Duluth, timber cruiser and surveyor, trained and inspired five of his seven sons to follow in his footsteps as expert woods observers. Confident that iron was to be found in the rugged ranges north of Duluth, the Merritt sons combed the forests with pick and shovel. In March of 1889 they turned up some red sandy material, heavy but soft and friable. The ore assayed at 60 percent iron, but they were nevertheless disappointed that what they found was not hard-rock ore like the Vermilion's. Soft, loose ore like this had never been seen before, let alone mined.

The Merritts' search went on, and a year and a half later, Captain J. A. Nichols, who was in the Merritts' employ, found more soft ore, a richer deposit yet, just west of the future town of Virginia. A small test pit suggested that great amounts might lie beneath the pine needles and thin glacial drift. The location immediately was given the appellation Mountain Iron; the date was November 16, 1890.

Soon, all around Mountain Iron, Merritt test pits yielded soft ore grading around 65 percent iron. The Merritts made no attempt to keep their find secret, and news spread like wildfire, in Duluth, across the country, and the rush was on. In less than two years, more than 100 mining companies were formed; the Merritts were involved in twenty of these ventures.

Realizing that profitable exploitation of the ore required transportation, the Merritts built the Duluth, Missabe & Northern Railway, which, connecting with the existing Duluth and Winnipeg Railroad at Stony Brook Junction (now Brookston), provided rapid transportation to docks in Superior, Wisconsin— and to shipping on the Great Lakes.

With loans readily available in an iron ore euphoria, they extended their capital investments rapidly, purchasing land, mining equipment, larger locomotives and ore cars, and extending their railroad to shipping docks in Duluth. The Merritts started with their own capital of $20,000 and at the peak of their mining activity turned down an offer of eight million for their interests.

In 1893, however, the country was hit by one of the worst economic depressions in history, and mining operations on the Mesabi came to a virtual standstill. Hopelessly in debt, the Merritts went broke. Expert woodsmen and romantic explorers—the Du Lhuts and Des Groseilliers of an era past—the Merritt brothers had not grasped the financial and political complexities of Wall Street and Washington. That they soon passed from the scene of big business in iron was, perhaps, inevitable.

The Merritts' interests and holdings were taken over by default by financier John D. Rockefeller, to whom they were deeply in debt, and whose monetary resources could tide over the Panic of 1893. Rockefeller's involvement—and later, the participation of Andrew Carnegie, James J. Hill, and J. Pierpont Morgan— eventually led to the formation of the United States Steel Corporation, business giant of iron ore mining, ultimately controlling three-fourths of the Mesabi's ore production, and most of the Vermilion's as well.

In subsequent years, more mines opened, thousands more miners invaded the Giants Range.

The soft ore, near the surface, merely had to be scooped up and hauled away; it was a far easier and more profitable job than drilling the deep, hard-rock underground shafts of the Vermilion. Scoooping up the ore was made even more rapid with the introduction of the first steam shovel, weighing thirty-five tons, in the winter of 1894. Existing railroads could carry the heavy machine only in winter, when the ground was solidly frozen. Open pits were dug larger and deeper, 500 feet and more, lined with terraces that allowed loaded trains to crawl their way over switchbacks on their way out of the pits; trains were later replaced by large diesel trucks. Larger and larger steam shovels were introduced, up to 350 tons, later replaced by huge electric shovels. More railroads were constructed, more firmly built, with bigger locomotives and more capacious cars.

By the end of the 1890s, Rockefeller's fleet of ore freighters numbered twenty-eight and, by 1901, more than twice that number, constituting a vital link in iron ore transportation between the mines north of Lake Superior and the blast furnaces of the lower lakes and eastern cities.

The Mesabi Iron Range, as it turned out, was 100 miles long, with a great Z-fold approximately in the center near present Virginia, the Virginia Horn. Ranging up to two miles wide, and in places almost pinching out altogether, the deposits led westward from Birch Lake and Babbitt through mining towns of Embarrass, Aurora, Biwabik, and Eveleth to the Z at Virginia; on westerly through Mountain Iron, Chisholm, Hibbing, Nashwauk, Calumet, and Coleraine, the range tapered out just west of Grand Rapids. These are Minnesota's "Range Cities."

Soon after the First World War, the largest and richest deposit of all, near "old" Hibbing, was found to extend under the town. And so Hibbing itself was moved, to allow extension of the great Hull-Rust-Mahoning pit, biggest in the world at over three miles long and 500 feet deep.

The Mesabi became the unchallenged world leader in production of iron ore. In the last decade of the nineteenth century, the Mesabi produced thirty million tons of high-grade ore.

Still more was to come. Beginning early in the twentieth century, the Mesabi produced over half of United States iron ore. Maximum production was achieved in the 1940s, when the demands of the Second World War were met with increased effort, totaling over a half-billion tons for the decade, almost all natural ore. In 1942 alone, the peak year, the Mesabi yielded seventy million tons. By 1960, the Mesabi high-grade ores were nearly exhausted, but they had reached an impressive total of over two billion tons.

In little more than a half-century, the legendary Mesabi, sleeping for two billion years along the slope of the Giants Range, gave up its greatest treasure. The soft, easily worked ore had been extracted in a frenzy of exploitation, railroad building, capital investment, North Shore navigation, and the greatest human immigration and settlement that the land north of Superior had seen yet.

The Mesabi had become the richest iron mountain in the world. Its ore had made and broken millionaires and plunged America's greatest financiers into the biggest of big business. It went on to provide the greatest source of iron and steel for the tools, weapons, and ammunition of the Second World War. And it left a dozen modern cities carved out of the wilderness, unmatched in their own treasure trove of iron ore revenue.

But, except for small pockets of remaining high-grade ore, the inexhaustible had become exhausted. Today, the McKinley Mine, near Aurora, survives on the Mesabi, the last sizable operation to mine natural ore. Since 1980, the McKinley has yielded about fifteen million tons. It is estimated to continue only through the 1980s.

Outside the Lake Superior basin, but related in its iron formation to the Vermilion and Mesabi ranges, was Minnesota's Cuyuna Range. The Cuyuna lay along the upper Mississippi River, but it contributed its iron to the main Lake Superior ore traffic. Opening in 1911 and largely closed by 1981, the Cuyuna produced more than 100 million tons of iron ore.


The soft ore of the Mesabi was high-grade, direct-shipment ore, easily mined, needing no mixing. But the Mesabi deposits had formed, not in broad, flat fields, but rather by an unpredictable natural beneficiation into relatively small deposits, surrounded by incalculable quantities of taconite. Edward W. Davis, minerals engineer at the University of Minnesota and pioneer of taconite processing, has described the rich Mesabi deposits as "raisins" in a "cake" of taconite.

Unusable in the steel-making industry because of its low iron content—about 30 percent—taconite is not considered "iron ore." Yet, with the Mesabi's natural ore nearly gone, taconite in its enormous quantities awaited only human technology to create another seemingly inexhaustible supply of iron.

The need for industrial beneficiation of low-grade taconite was recognized early in the twentieth century. The low-grade materials found by early prospectors prompted many attempts involving washing, crushing, sifting, heating, and other such treatment, and some were moderately successful, like sintering. Even at a low grade of 30 percent, the superabundance of the formation meant huge quantities of potential iron, if a means could be found to upgrade the taconite rock to iron concentrations and physical shapes economically useful in the industry's smelting processes.

Edward Davis began work at the university's Mines Experiment Station in 1913 and almost immediately began to attack the problem of beneficiation of taconite. He was to spend forty years in his quest.

A test mine and plant were opened in 1918 on the eastern Mesabi, in the area of Christian Wieland's first discovery of magnetic ore. The Mesabi Iron Company was formed; several thousand tons of beneficiated taconite were shipped. A mining town was established nearby with the name of Argo—soon changed to Babbitt. But the process was too expensive to compete with natural ores elsewhere, and Davis went back to the laboratory.

A key element in taconite processing is that the iron occurs in tiny black grains that are magnetic. After loosening by crushing and grinding, the grains—high-grade magnetite—can be extracted with strong magnets. Small wonder that Christian Wieland's compass had gone astray.

Davis and his associates eventually did perfect the pelletized taconite product of today. The process developed into one of crushing and grinding, washing, magnetic separation of the high-grade particles, and agglomeration into marble-sized balls, using as a binding agent the volcanic clay bentonite. The final product graded 63 percent iron. The spherical shape was a major breakthrough, too, for it allowed the maximum interstitial spaces between pellets, necessary for the flow of air in blast furnaces. The pellets performed even better than natural ores.

In 1939, the newly formed Reserve Mining Company acquired the Mesabi Iron Company and its holdings, including the old test mine near Babbitt. Taconite processing required immense quantities of water, and since about two-thirds of the mined taconite ends up as waste, or tailings, great quantities of unwanted material were to be produced. Therefore, when selecting a site for a processing plant, Reserve Mining Company picked a location on the shores of Lake Superior, where water could be easily obtained and tailings disposed. The processing plant was built in the mid-1950s, and Reserve Mining began its large commercial production of high-grade pellets. The town of Silver Bay, largely built by the company, blossomed at the site. And soon the production of iron from taconite pellets surpassed that from the Mesabi natural ores. Reserve's plant accounted for 17 to 18 percent of United States steel production. Eventually, eight taconite mines and plants extracted and processed the "cake" from the Mesabi, Reserve's being the only plant located on the shore of Lake Superior. The tailings disposed of so easily into Lake Superior later were found to present an environmental and health threat, and the ensuing controversy occupied the courts and people of Minnesota for several years in the mid-1970s, until the disposal of tailings was ordered on land.

In the early 1980s, a depression in the American Steel industry greatly reduced the need for Minnesota's iron ore. But by this time, total production from the Mesabi had reached the impressive level of three and one-half billion tons, including both natural ore and processed taconite pellets.

Now the rush is on again. Once more, the magic call of gold flashes across the continent, and the fever is up to a new pitch.

A few years ago, Hemlo, Ontario, was a railway stop and cottage settlement located at the north end of Cache Lake, east of Heron Bay, along the tracks of the Canadian Pacific Railway. Terrace Bay and Marathon were young, progressing pulp-and-paper mill towns; Heron Bay served the soon-to-be-opened Pukaskwa

National Park, Canada's newest. Today, the area is abustle with huge trucks, bulldozers, construction of head frames and machinery sheds—all because of the discovery of gold. One major deposit even lay beneath Trans-Canada Highway No. 17, the North Shore Drive. Three main mines are all located within sight of the highway.

This gold rush does not now comprise the lonely prospector with pick and shovel, canoe and pack, however; nor is it a stampede of amateurs arriving by steamboat, horse, and huffing steam railroad. It is now systematic industry—replete with geologists, engineers, metallurgists, chemists, computer technicians, and well-financed and experienced mining companies. Helicopters, not canoes, now penetrate remote areas.

Gold had been known in the Hemlo-Heron Bay area for many years. As early as 1869, gold-bearing veins were discovered and several shafts sunk near Heron Bay, and some shipments of ore were made. William Frue, of Silver Islet fame, was involved. In the decades of the first half of the twentieth century, pits, trenches, and test drill holes were made intermittently. Then the exploration stopped—for a while.

In 1947, the first large, modern staking was made by Lake Superior Mining Corporation, and in 1981 International Corona Resources, a major holder, announced a significant deposit. The new rush was on. Soon, 7,000 claims were staked, involving 40 to 50 companies. By 1983, an estimated 16,000 claims had been staked, and more than 150 companies were active in this northeastern corner of the North Shore. Now three major mines are operating: Lac Minerals, Noranda, and Teck-Corona.

Rather than searching for placer deposits eroded from surrounding hills into the gravel bars of valleys and streams, this new enterprise seeks out native gold in hard-rock strata, deep below the surface. Deep drilling is employed. Nor does the gold exist as veins in volcanic lavas, as is the usual case elsewhere. Here minute particles, invisible to human vision, were incorporated in volcanic ash of Archaean age, later hardened to sedimentary rock. The ore grades around a quarter of an ounce of gold per ton, and great volumes of the sedimentary rock are present. A day's production: about forty pounds of gold.

The full story is not in yet, but total dollar worth is estimated at ten billion, currently one of the largest strikes in the world, the biggest gold rush in Canadian history.

SIX

# *The Tall Pines*

etween northern conifers and the deciduous hardwood trees to the south, a belt of mixed forest extended in east-central North America from the Maritime provinces and Maine, southwest up the St. Lawrence Valley, to cover most of the Great Lakes region. Here a great diversity of tree species created a rich woodland mosaic.

This mixture had developed because the coniferous and deciduous forest types had approached each other, from north and south, across a climatic tension zone—an area that fluctuated in its forest composition as climates fluctuated north and south following glacial retreats. Lake Superior was itself a major factor in holding back the Arctic influence. We know these woodlands with mixed species today as the Great Lakes-St. Lawrence Forest Region.

A bit of the southern deciduous forest still wraps around lower Lake Michigan and the easternmost of the Great Lakes, Erie and Ontario; a bit of the northern coniferous forest—the Boreal Forest Region—comes down to wrap around part of Lake Superior's northern coast. Otherwise, the mixed Great Lakes-St. Lawrence Forest extends in a wide, irregularly shaped band from the Atlantic Ocean to the western head of the Great Lakes. It contains representatives from both the boreal and deciduous forests:

*Red pine and white pine*

conifers such as pines, spruces, and balsam fir, hardwoods such as sugar maple and red oak, aspen and birch.

In this zone of mixed forest species the white pine reached its preeminent position. And scattered with the white pine grew also the red, or Norway, pine, especially in northern parts of the zone. In the history of the Great Lakes lumber industry, the white and red pines were known as the *tall pines*—distinguishing these two from the lesser jack pine, which had coarse and weak wood. Crowded close together, the tall pines grew to immense heights to reach the sun. Among the trunks, where even the lowermost branches were far overhead, only shadows fell upon the ground, and only browned needles covered the spongy earth.

Owing to fires in prehistoric times, pure stands of mature pines occurred in a mosaic, mixed with immature stands or other species. In the pure stands, separately covering hundreds of square miles, the tall pines grew in a profusion of gigantic trees, in darkened woodlands that seemed to have no end. Towering high and straight-grained, without branches or knots, the huge trunks invited the logger's ax.

On the Atlantic coast the white pine early provided masts for England's navy, later squared logs for England's sawmills. At the first water-powered sawmill in the upper Great Lakes area, pine was cut in 1783 at Sault Ste. Marie for the posts and boats of the North West Company's Lake Superior fur trade, a mill that was destroyed a few years later by American forces in the War of 1812. But a century later, white and red pine from the upper Great Lakes region had provided lumber to build the major cities of the Canadian and American Midwest.

The tall pines spawned the lumberjack era, a century that still echoes the ring of the faller's ax and the rumble of brawling river drives. It was as lusty and colorful as the fur trade, the mining, and railroad-building periods.

An end came, however, to the era of the tall pines. And when it did—to the apparent great surprise of the lumber industry—its demise left vast areas of incredible desolation: sere forests of huge stumps, withered slash, eroded slopes, and great fires.

On the cutover lands the white pine did not regenerate. The big trees had been cleared out so thoroughly that in many large areas too few mature trees remained to provide seed for natural reproduction. An even more important factor, however, was the inadvertent introduction of blister rust from Europe; this disease not only prevented natural regeneration but also frustrated our own attempts at successful artificial propagation.

So northern conifers and aspens recolonized the empty expanses that once held such great stands of white and red pine. The shifts of forest succession in the tension zone continue, now with added human factors causing so drastic an effect.

To eastern and midwestern lumbermen, no other tree could match the white pine. Its wood was strong, light and buoyant, straight-grained, and durable. In fact, to these lumbermen, the term *lumbering* meant *white pine*. From the Maritimes to Minnesota, the white pine was the predominant lumber species. It grew to great sizes—up to 150, even 200, feet tall and up to 5, perhaps 6, feet in diameter—in a period of 400 to 500 hundred years. Large trees contained 2,500 board feet of lumber, the largest up to 6,000—single logs up to 500 or more.

The white pine's abundance elicited many statements by leading politicians and businessmen to the effect that its natural supply was inexhaustible. But after the great pine forests in east-central Minnesota, along the Mississippi and St. Croix rivers, were cut in the latter half of the 1800s, the supply was approaching exhaustion. Lumbermen's eyes turned north, around Lake Superior.

A decade after the turn of the century, the continent's last large stands of white pine remained in Minnesota and western Ontario, from Lake Superior north to the border. In the northern part of this area, around the maze of border lakes, were the richest stands. Only small tracts were to remain uncut.

Along the North Shore white pine had not developed large trees, like those to the south and east, or dense forests, like those along the border lakes. In fact, the distribution of white pine along the North Shore extended eastward from Duluth only to Thunder Bay, north of the international border, with boreal forest farther east and north.

Some red pine, too, grew north of Superior, more successful on the lighter and rockier soils. Unlike the white pine, the red pine is more demanding in its environmental requirements and therefore less adaptable. Requiring open sun, it could not compete with brush or overstory, and thus was limited largely to islands and lakeshore where full sunlight was available.

With the pine gone in the late 1800s from the rest of the Great Lakes forests, and eastern markets beyond the locks at Sault Ste. Marie (opened in 1855) still clamoring for lumber, northern Minnesota pineries remained attractive. Mills in Duluth waited to convert logs to lumber, and steamers plying the waters of Lake Superior waited to carry sawn lumber to eastern markets.

A major problem, however, was soon encountered on the North Shore.

Up to now, most of the pine harvest in the Great Lakes forest had depended upon the river drive. Logs were cut in winter, skidded by oxen or horses to be piled along major streams until spring, and then driven down with the high water of snowmelt. It was rare, indeed, that major tracts of pine were so isolated from rivers that they could not be cut and transported by water.

But the streams on the Minnesota North Shore were not conducive to the river drive. Entering Lake Superior along a steep, rocky coastline, North Shore streams with rare exceptions all dropped precipitously in waterfalls and wild cascades—a total fall of 500 feet or more in their last plunges to the lake. The loss of logs through breakage and jams would have been prohibitive.

Nevertheless, logging commenced on the Minnesota North Shore in the 1880s, and although limited to areas near the lake, the lumber business boomed. Pine logs were cut, landed directly on Lake Superior's shores, and then rafted to several huge sawmills recently built at Duluth.

Meanwhile, the pine harvest from inland regions awaited the development of the next log transportation system that followed the river drive: the logging railroad.

By the late 1880s, harvest of Minnesota's last large stands of tall pines was well under way, much of the cutting along the North Shore. The big logs were skidded out with animal power, or sleigh-hauled by horse in winter, to Lake Superior shores. Cutting was done only within a maximum distance of ten miles from shore, usually less. And the sawmills, idle in winter, were supplied only after spring rafting began on Lake Superior. But when the new technology of logging railroads arrived, this

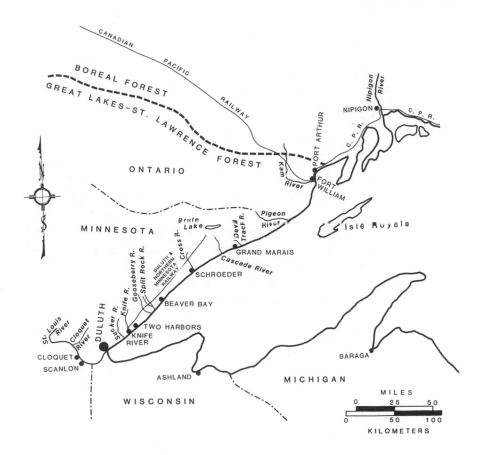

*Logging of the tall pines—white and red—in much of the Great
Lakes-St. Lawrence Forest of mixed conifer and hardwood trees between Duluth and Fort William/Port Arthur. Many small logging
railroads carried pine logs to mills on Lake Superior.*

scene changed quickly. The earliest logging railroad in Minnesota
appeared in 1886.

At first, the small railroad spur merely substituted for the
short haul by horse, from cutting ground near rivers or
lakeshore. By 1900, however, short spurs stitched the pine woods
above Lake Superior's shores with hundreds, later thousands, of
miles of steel tracks. The skein of rails probed into the roughest
country to take out otherwise unavailable pine. Logs were railed
down to landings on the lake shore—at Knife River, Two Harbors, Gooseberry River, Split Rock River, Beaver Bay—and made

up into great rafts, or booms, containing millions of board feet, for towing by steam tugs to mills at Duluth, or Ashland, Wisconsin, or Baraga, Michigan. The record raft on Lake Superior contained six million board feet, in a tow from Two Harbors to Baraga, more than 200 miles, in 1901, a particularly long and treacherous tow.

Towed "rafts" on the Great Lakes were not like the rigid, chained or roped rectangular rafts on the Mississippi River. The river raft was tried as early as 1857 on Lake Huron, but the huge swells and waves of the big lake would break up these rafts, and losses discouraged rafting for a while. Later, the lake raft was developed, consisting of great numbers of free-floating saw logs contained in a huge boom of larger logs, chained end to end. In 1885, the first of the lake rafts, containing three million board feet of white pine, was towed across Lake Superior, driven down the St. Marys River, and towed down Lake Huron to Bay City, Michigan. Because of storms and swells on the lakes, large rafts were more successful than small ones. In the 1890s decade, more than two billion board feet of lumber were sawn in Duluth and Wisconsin mills from lake-rafted logs.

Later, main railroad lines carried pine from Superior highlands directly to mills at Cloquet and Duluth. About 1900, railroads were moving three million board feet in logs daily to Duluth mills; and in the peak year—1901—more than a billion board feet in logs were railroaded into Duluth. A major advantage in the use of logging railroads was that their year-round use allowed the mills to operate year-round, too, instead of lying idle for much of the winter, waiting for their supply from spring river drives or rafts from the lake.

The era of temporary logging railroads developed its own mixture of problems and jargon. For example, laying of spurs across swamps presented special difficulties: tracks would sometimes founder in sink holes, and locomotives as well as rails, cars, and logs were lost in the muck. "Skeleton" tracks—ties and rails only, without ballast—were laid on frozen ground to be used only through winter; but these were subject to "sun kinks" when spring weather warmed and expanded the rails, causing derailment and abandonment of the line for the season. When tracks ran down steep hills, and loaded trains mainly went one way— down—the rails had a tendency to move downhill, too, causing kinks and derailments. Forest fires, caused by sparks and live coals from locomotives, were common until rigid controls were implemented.

However, despite these operational problems and the twisting nature of the lines, the logging railroads quickly completed the job of removal of white and red pine from the Minnesota North Shore. By the mid-1920s, most of the logging railroads were gone from northern Minnesota. The rails were taken up for salvage or installed elsewhere.

The era was short—only thirty years. But in Minnesota as a whole, forty companies and about 5,000 miles of logging lines, mostly north of Superior, were involved in this last major tall pine harvest.

Many lumber companies operated to remove the pine logs. Some were large and influential and persisted through to the present day; some were small and short-lived. Some, although operating on Minnesota's North Shore, were based out of state; for example, several Michigan and Wisconsin companies were active in northern Minnesota after the pine was gone from those two states. Some companies incorporated their own logging railroads; others, smaller, used the services of independent logging railroads or main lines.

On the Minnesota North Shore, the Alger-Smith Lumber Company stands out as the largest operator; a Michigan-based firm, it was chartered and set up headquarters in Duluth in 1898. The company included the Duluth & Northern Minnesota Railway, the most important logging railroad in Minnesota. Its tracks extended northeast to near Brule Lake, and they brought white and red pine logs to the company's main yard at Knife River. Alger-Smith also cut all the way up in the Pigeon River country, along the international border, and rafted down the lake shore to Duluth—with booms containing up to five million board feet. The company logged the North Shore for about twenty years, finishing by 1919; the last of Alger-Smith's rails were taken up in 1923.

Other operations on the North Shore included the Split Rock Lumber Company and its Split Rock & Northern Railroad, which worked over the country around the mouth of the Split

Rock River from 1899 to 1906. The steam tug *Gladiator* towed rafts to Duluth and took along carrier pigeons to send distress messages if necessary.

The "Estate of Thomas Nestor," a Michigan lumber firm, logged the Gooseberry River area with the company's own railroad equipment from 1900 to 1909, rafting logs to Ashland, Wisconsin, and Baraga, Michigan.

The Brooks-Scanlon Lumber Company cut the record section of pine in Minnesota—thirty-three million board feet—along the headwaters of the Sucker River, the famous Section "35-53-13." Brooks-Scanlon's logs were railed by its Minnesota & North Wisconsin Railroad to mills in Scanlon, a new sawmill town north of Duluth, operating from 1901 to 1909.

Smaller lumber companies and their associated railroads logged off other small portions of the shore. Tracks were laid inland to haul logs from otherwise inaccessible areas to the lake, then rafts were towed to the mills at Duluth or, in some cases, to Wisconsin mills. Other companies railed logs directly to Duluth.

One major logging operation that did use the traditional river drive, however, was that of the Schroeder Lumber Company on the Cross River. For a ten-year period—1896 to 1905—Schroeder cut logs on thirty-six sections of pine land in the drainage of the Cross. Schroeder constructed a wagon tote road up along the rugged course of the river and operated a series of driving dams in the spring, down a torrential series of falls and cascades. When logs would fill the pool impounded behind the upper dam, stop logs were pulled in the dam to release water and logs to the next pool, and the operation was repeated through the next and succeeding dams until that lot of logs was landed on Lake Superior. Later, after all pools were full of water again, another drive of logs was let down. Even so, the battering of logs caused fraying—or "brooming"—at the ends, and to compensate for this loss the logs were cut in the woods one foot longer than standard lengths, to be trimmed later. Schroeder's logs were boomed on Lake Superior and towed to mills in Ashland, Wisconsin.

Another unusual river-drive operation was that in the Pigeon River watershed, on both sides of the Minnesota-Ontario border. The lower Pigeon is noted for its succession of cascades and waterfalls, culminating in 100-foot High Falls. But a long flume was constructed around this last falls, and many were the white and red pine sawlogs that were cut in the upper drainage, on both sides of the border river, and driven down and around the falls in the flume. Many logs were broken or damaged in the Pigeon River drives, resulting in a unique pine lumber grade, No. 6, which was marketed to settlers at low prices. From 1899 to 1902, Alger-Smith cut timber in this area and towed pine logs to its mills in Duluth. Later, the Pigeon River Lumber Company, a Wisconsin firm, logged the remainder of the drainage and towed logs to mills in Port Arthur, again from both sides of the border. The Port Arthur mill closed in 1919, when the pine, after twenty years of cutting, was gone from the Pigeon River.

Farther south, Andrew Hedstrom, an immigrant from Sweden, in 1914 set up a commercial sawmill on the banks of the Devil Track River and cut saw timber from surrounding pine woods, hauling logs from camp to mill by horse and sleigh. In addition to lumber, the Hedstrom Lumber Company produced shingles, railroad ties, and fish boxes. These products all had to be hauled from the mill at the top of Maple Hill—now part of the Gunflint Trail—down to Grand Marais, and all supplies *up* Maple Hill. Hedstroms still operate the sawmill on the banks of the Devil Track, providing lumber and other wood products to the town of Grand Marais and surrounding areas.

The mills at Cloquet first received pine logs driven down the Cloquet and St. Louis rivers; later logs were transported by railroad. These mills at Cloquet developed where they did because the torrential lower gorge of the St. Louis River prevented log-driving down to Duluth. The great fire at Cloquet in 1918 burned the town completely, including sawmills, and took more than 500 lives. By this time, however, the pine harvest was nearly complete, and the pulpwood era was beginning. Cloquet was about to emerge, Phoenixlike, to become a major paper mill town. Today, the Potlatch Corporation, part of the huge Weyerhaeuser empire, manufactures some of the highest grades of fine paper at Cloquet.

One more pine operation remained to be carried out. The General Logging Company and the Duluth & Northeastern Railroad cut pine in the Cascade River area and railed its logs to

Cloquet, a distance of 131 miles. It was a busy line, with twelve trains running daily each way. It was the last of the great tall pine logging on the North Shore, operating only for the two years 1927-28.

Today, in addition to pine stumps and scattered, still-resistant pine knots (favored for producing hot fires in backcountry wood ranges), the occasional abandoned logging railroad grade can be rediscovered by the wandering grouse hunter or trout angler. Some of the main lines were continued as more permanent railways; some were converted to forest truck roads. Some are now used as hunting, skiing, or snowmobile trails. The overgrown grades, and occasional rusting rails and artifacts from the old camps, are reminders of this colorful period in tall pine logging along the North Shore.

Logging of white and red pine on the Ontario east side of Lake Superior commenced only a little later and in the same fashion as in the west, although on a less extensive scale.

In eastern Canada, lumbermen had cut the pine and moved west; pineries of the Ottawa River country included some of the richest stands of white pine in Canada. Along Georgian Bay, then up along the North Channel above Lake Huron, marched the advancing line of logging camps in the latter half of the 1800s, toward Sault Ste. Marie at the outlet of Lake Superior.

But here abundant pine timber stopped. Along Lake Superior's eastern shore, part of the Great Lakes-St. Lawrence Forest, red and white pine grew only in sparse density, less abundant than on the northwest coast in Minnesota. The main source of pine logs in western Ontario was not along the North Shore, but rather in stands of the Rainy Lake and Lake of the Woods regions to the northwest. Since this area lay over the watershed divide in the Hudson Bay drainage, driving rivers did not lead to existing mills. Consequently, this last stand of pine remained for the railroad to reach it and transport logs out.

The mills in Fort William and Port Arthur were hungry for pine in 1885, and when in that year the Canadian Pacific Railway was completed, one of its more important functions was to transport pine sawlogs. A second important function was to carry pine lumber west to the developing prairie settlements in Manitoba.

To the southeast, north of the Sault, lay the great, sweeping hills of old granite mountains, in the Algoma District. Much of the terrain consisted of east-west running ridges; on the more gentle northern slopes were vast sugar maple woodlands, while steeper southern slopes held some white and red pines, the object of the last lumberman's eye.

Chartered in 1899, the Algoma Central & Hudson Bay Railway in the first few years of the new century was being constructed with great difficulty through the extremely rough Algoma country, north from Sault Ste. Marie. The principal purpose of the railroad was to furnish Francis Clergue's lumber and steel enterprise at the Sault with logs and iron ore, but it was also intended to assist in colonization of the district—the elimination of the forest being one of the prerequisites to settlement. By 1903, when an economic crash brought a temporary halt to the railway's construction, rails as far north as Mile 56 had been bringing some pine sawlogs to mills at Sault Ste. Marie. Logging for the tall pines continued at a low level up to the 1930s, with a peak of activity around 1910.

The Algoma Central never made it to Hudson Bay, but it was extended considerably farther north in the early 1900s, to an important connection with the Canadian Pacific that wound eastward from Fort William and Port Arthur along the North Shore.

The Algoma Central and the Canadian Pacific were to figure prominently in the next phase of forest exploitation along the North Shore—the harvest of a vast resource of black spruce.

Not for sawlogs of the tall pines now, nor lumber to build cities. Rather, the logging for spruce pulpwood was to feed another of the continent's developing appetites: for paper and a burgeoning diversity of paper products.

North of Sault Ste. Marie, from approximately Michipicoten Harbour on the eastern shore westward around the lake to the Pigeon River, grew a dense forest of northern conifers, dominated by the black spruce. This was the Boreal Forest Region, extending north in its vastness from Lake Superior to meet the Arctic tundra—a tree line that had followed retreating glaciers, sweeping northward across the Great Lakes region thousands of years ago.

# SEVEN

# The Boreal Forest

In a small portion of the Great Lakes region, a bit of the northern coastline on the northernmost lake, there grew a part of the northern Boreal Forest with its dense legions of black and white spruce. Here the spruces had fought for domination with the tall pines several times since ice covered the Great Lakes region. Now, in the early decades of the twentieth century, with the pines essentially gone as a result of logging at either end of Lake Superior's North Shore, the spruces of the Boreal Forest Region beckoned new loggers and mills that turned toward another enterprise growing from the pines' ashes.

The new industry rose to serve the clamoring demands of a maturing and increasingly complex society—for newsprint, paper for books and letters and business records, for an increasing variety of paper products.

And so new lumberjacks cut these lesser trees of the boreal forest: the spruces primarily, and associates like balsam fir, jack pine, and tamarack, for pulpwood and the production of paper.

The first major logging of timber in the boreal forest, however, was for neither pine logs nor spruce for paper. Rather, earliest logging was related to the needs of railroads in the 1870s and 1880s—for ties, bridge and trestle timbers, and logs for pilings in wharves and piers at lakeshore shipping facilities.

Cutting grounds and sawmills were located only close to rail lines, and not until after the Second World War, about 1948, did trucking open up more extensive forests to either pine or pulpwood logging. In fact, it was the railroad that in the early years most influenced logging objectives and cutting methods in the Lakehead region: primarily the Canadian Pacific Railway, and also the Canadian Northern (now Canadian National) and, later, the colorful "Pee Dee" (Port Arthur, Duluth, and Western Railway).

This last, the "PD&W," was otherwise and disdainfully dubbed the "Poverty, Depression, and Want." Its rails never reached Duluth but for a while served the area between Port Arthur and the border lake country with a successful trip being one in which the train went off the track only two or three times.

Logs for pilings were cut along shores and islands and towed on Lake Superior, transported by railroad, or hauled by sleigh in winter. Pile logs were cut in 1873 for the first Federal Government dock at Prince Arthur's Landing, on Thunder Bay. At first tamarack, resistant to rotting, was the favored species for pilings; after vast tamarack stands were eliminated in the late 1890s by the pestilential larch sawfly, jack pine and spruces were used. Pile logs ran up to seventy-five feet long, averaged forty feet. Between 1870 and 1930, one million pile logs were cut, equivalent to 200 million board feet.

Logging for pile logs, however, was a tiny enterprise compared with the production of railroad ties, for which the developing railroads had ravenous appetites. At first, hemlock ties were imported from Georgian Bay on Lake Huron and landed at Heron Bay and Rossport, where the Canadian Pacific Railway was under construction. But this supply was insufficient, and soon vast quantities of tamarack (prior to the sawfly) and jack pine were cut for ties and bridge timbers. The railroads' use of ties and timbers was not only in the North Shore region of Ontario, but also in extension of tracks into the newly settled prairies of Manitoba.

The manufacture of ties was an art learned in the bush, requiring skilled men: logs were squared by hand-hewing with a

*Pulpwood boom, Pic River*

broadax. Working on the cutting ground, a good axman made thirty to fifty ties per day. Tamarack was preferred; jack pine was a good second choice. Ties were sold in three grades, and a top-grade tamarack tie was sold by the logging company—cut, hewn, hauled, and delivered—to the railroad, for about twenty-five cents each, jack pine a few cents less.

Between the town of Chapleau on the east side of Lake Superior, and the western Ontario-Manitoba border, from 1875 to 1930, fifty-five million ties were produced, equivalent to two billion board feet. Most of these North Shore ties were used in laying tracks to the western prairies.

The largest tie-manufacturing operation in the Superior region, providing by far the single greatest quantity of ties for the Canadian Pacific Railway, was that of the Austin Nicholson Company. Based in the Chapleau district, the company made ties that supported the engines and cars of the Canadian

Pacific literally around the North Shore, from Chapleau north and west to Manitoba.

The Austin Nicholson Company began in 1901 at Chapleau along the Canadian Pacific main line. Founders James M. Austin and George B. Nicholson soon established major operations at Nicholson's Siding (later the townsite of Nicholson), a few miles northwest of Chapleau on the Canadian Pacific tracks and along the shores of Windemere Lake. Bush camps were established all around the lake and among a complex of connecting rivers and lakes (which eventually drain by way of the Shikwamkwa River to the Michipicoten River and Lake Superior). Supplies were transported to the camps across water, or across ice in winter, while logs and ties were floated or sleighed to mills at Nicholson's Siding. The Windemere Lake area was rich with great stands of tamarack and jack pine.

In addition to the Windemere Lake area, the company's operations ranged along the Canadian Pacific tracks from Chapleau to Heron Bay, totaling in its first decade more than 1,000 square miles of cutting rights.

Austin Nicholson was noted for introducing the mill-sawn tie—rather than ax-hewn in the bush—a product much more efficiently produced, with the added advantage of using slabs in the mill for production of lath.

A new mill was built at Nicholson's Siding in 1915, and tie production reached 2,000 ties per ten-hour shift, with 18,000 pieces of lath as a by-product. A normal day's operation was two ten-hour shifts. By 1916, Austin Nicholson had become the largest producer of ties in Canada, a position it maintained for over a decade. The company had a major impact upon the economy of north-central Ontario, and George Nicholson became a central figure in the social and political arenas of the region.

In 1918, the town of Nicholson was established, a company town with family homes, school, churches, poolroom, and ice cream parlor. In 1921, a new, large mill (and eventually the town of Dalton) was established farther up the tracks along the Shikwamkwa River system. Other mills and company towns were established farther east.

The demand for ties reached a peak in the 1920s and dropped thereafter. A major factor in the decline was the introduction of creosote, a wood preservative. Without preservative, the average life of a jack pine tie was about seven years; with creosote, a tie's life was extended to fifteen to twenty years. By 1922, all Austin Nicholson ties were creosote treated.

James Austin died in 1922, and the Nicholson mill was destroyed by fire in 1931. By that time the great Windemere jack pine stands were gone, and, furthermore, the depression had brought a virtual standstill to business.

And so the company was dissolved in 1934. The Austin Nicholson empire had logged about 2,000 square miles of jack pine forests; it had introduced the mill-sawn tie, of which it produced millions for the Canadian Pacific Railway. And it left its ghost towns along the shores of Windemere and the Shikwamkwa.


As the pine ran out along with the nineteenth century, old camps had been dismantled, logging rails taken up, and the tall-pine lumberjacks had moved west. In their place came a new breed—of camps, transportation, and men. The old log bunkhouses were replaced by housing more like dormitories; animal-drawn sleigh hauls were long gone and now of historical interest, and transportation of logs depended less and less upon the river drive, all to be replaced by trucks. And the men were no longer the rough-hewn woodsmen of the past but skilled technicians and heavy equipment operators.

The passing of pine logging and the lumbering industry, and their replacement by the pulp and paper business, seemed to fit appropriately with changing events during the period of white settlement of the continent. The pine had provided lumber for the building of homes and factories and schools, lasting long enough for the completion of major Midwest and prairie cities; the new civilization now was ready for proceeds of the remainder of the northern woodlands—pulpwood for paper and paper products.

The trees, too, were new breeds. The first species that were the object of the pulp cutter were the spruces—black and white—with their naturally long fiber that was such a great advantage in the paper-making technology of the time. And at first, only the spruces were taken out.

In the Ontario Lakehead region, in the boreal forest which

had not been previously logged because pine occurred there so sparsely, it was the black spruce that predominated and which attracted the interest of the paper industry. The black spruce was the smaller of the two, rarely reaching eighty feet tall and two feet in diameter; the more usual size cut was thirty to fifty feet high and less than a foot in diameter. It flourished in great abundance, however, especially in swamps and wet lowlands.

The white spruce was common, though less abundant than the black, and larger; often as tall as a hundred feet and two to three feet in diameter, the white spruce was stately and graceful. If the white pine was king of the forest, the white spruce must rate the appellation of queen—surrounded by her endless minions of dark and glowering black spruce.

While pulp logging emphasized the spruces in the boreal forest, the pine cutover lands at either end of the North Shore sprouted other species, such as jack pine, balsam fir, tamarack, which began to be utilized by the paper industry as its technology improved. Still later, the aspen was to enter the paper industry as an important pulp species.

The mills that used pulpwood for paper also utilized these secondary trees for other products: aspen for lumber, excelsior, wafer board; birch for veneer in the furniture industry, for boxes and containers, matches; cedar, resistant to rotting, for fence posts and poles, logs for cabins and houses, mine timbers. White spruce, in addition to paper pulp, was also sawn for lumber, and highly prized at that.

Although some primitive paper-making operations began in the Lake Superior region as early as the late 1800s, it was not until well into the twentieth century and after the decline of the tall pine cutting that the paper business started to increase greatly. In the 1920s, the industry boomed along the North Shore.

The pulp and paper industry was a different kind of operation than the "cut out and get out" system of pine logging, and it necessitated new perspectives and policies in the use of forest lands.

In the lumbering operation, sawmills that simply converted logs to dimension lumber followed the retreating line of pine forests. But paper making involved a more permanent industrial base, with greater capital investment, dependent upon a steady and reliable supply of raw material. Thus it was that the paper industry, instead of mining the woods to exhaustion and moving on, necessarily began to manage the forest on a long-term, permanent basis, with systematic cutting and replanting, nurturing new growth—sustained-yield foresty, it was called.

Indeed, it was this new resource philosophy brought about by the paper industry that initiated the foresty profession and the new "conservation movement."

Along the Canadian portion of the North Shore, modern logging for pulpwood began near Sault Ste. Marie with the Algoma Central Railway and the industrial empire initiated by Francis Clergue. Pulpwood harvest moved north and west—in step, essentially, with development of transportation mode and harvesting technology.

Clergue and partner Edward V. Douglas, both Philadelphia financiers, formed the Sault Ste. Marie Pulp and Paper Company in 1894, at the same time obtaining a concession for timber cutting rights on fifty square miles of Crown lands from the Province of Ontario. It was a landmark transaction that set in motion the giant industrial development at the Sault, and the pulpwood logging of Canada's North Shore.

This original concession specified that cutting could be done for spruce, tamarack, jack pine, and poplar—but not pine—on land within three and one-half miles of Lake Superior, and within the same distance from one or more rivers draining into the lake, west of Sault Ste. Marie. Obviously, the intent was *north* of the Sault, thence west. Taken literally, directly *west* of the Sault would have placed the loggers in pretty deep water or, if a little farther, in the state of Michigan.

At any rate, some open-ended clauses—the merits of which they had no doubt learned about in Philadelphia—permitted the partners to extend this concession, almost at will, eventually up to the border of the Pic River watershed, more than halfway around Ontario's part of the North Shore. In 1896, the company began producing groundwood pulp at Sault Ste. Marie.

Francis Clergue passed from the scene in 1903, when a financial crash wiped out, temporarily, the great Sault industrial complex. But in 1911, the company, having been taken over by Lake Superior Paper Company, entered into lease agreements with the Algoma Central Railway. These agreements allowed pulpwood cutting by the company on railroad land grants, promised by the Ontario government. The lease was for ninety-nine years and is still in force. With the original concession along lakeshore and rivers, and the Algoma Central cutting leases, the Lake Superior Paper Company began its march up the North Shore.

Rivers along the eastern shore, while rugged enough, still presented good driving water for logs, in contrast to the Minnesota coast, where streams entered Lake Superior with little or no driving water. And through the old granite mountains of the eastern shore, rivers sometimes tumbled in fascinating falls and chutes, but above falls the track of many rivers wound headward in sometimes intricate mazes of lakes and tributaries—surrounded by immense stands of spruce and jack pine. This meant that bush camps could be established and serviced *by water*, rather than by tote road. Downstream, rivers were improved for driving by blasting out offending waterfalls and rocky gorges.

Near Sault Ste. Marie, pulpwood cutting proceeded with river drives on the Goulais, Batchawana, and Chippewa rivers, commencing in 1913, by the new Lake Superior Paper Company. Operations moved north to the Agawa, Michipicoten, and Magpie rivers. The drives through Agawa Canyon were rugged and hazardous.

An ambitious proposal for "improvement" of the rough-and-tumble Sand River was put forth in 1927, including plans for much blasting of falls and chutes, clearing of river channel, and construction of piers at the mouth. The plan was not implemented, however, because of its high cost. Major drives on the Sand River were not conducted, and the scenic canyon-and-falls stretches near the river's mouth, now in Lake Superior Provincial Park, were preserved.

In addition to exercising its cutting rights along river and lakeshore, the company at this time was cutting on its leases along the Algoma Central, and the railway brought in large numbers of pulpwood sticks to Sault Ste. Marie mills. Cutting continued into the 1920s, later under the Abitibi Power and Paper Company name, which had taken over from Lake Superior Paper.

Logging along the eastern shore rivers in the 1920s and early 1930s was intense, although operations were reduced in the

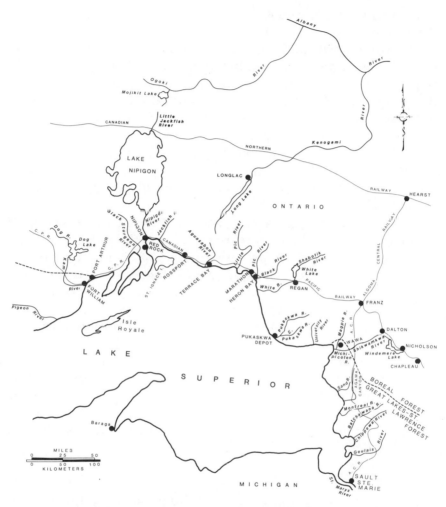

*Logging of black spruce and other conifers for pulpwood, mainly in the Boreal Forest north and east of Lake Superior. After cutting, the pulpwood "sticks" were transported by railway and river drive to paper mills.*

later 1930s because of the depression. Jack pine continued to be cut for ties, some white pine for boom logs, some of the larger white spruce for lumber. But by far the major cutting was of black spruce, for pulpwood. And in these two decades, the rivers on the eastern shore poured immense quantities of black spruce logs into Lake Superior, to be towed by steam tug to the company's mills at Sault Ste. Marie. By the late 1930s, the big river drives on the eastern shore were over.

Farther north and west, the forests of the wild Pukaskwa peninsula were also being exploited. Some early logging for white pine was undertaken near the mouth of the Pukaskwa River, in the period 1904-08. And pine logs were rafted to lumber mills at the Sault. Starting in 1917, Lake Superior Paper, exercising its old concession, began a much larger operation in the Pukaskwa watershed, now for pulpwood. With winter camps upstream on the main Pukaskwa River and on its major tributary, the East Pukaskwa River, spring drives to Lake Superior brought out huge quantities of spruce.

The Pukaskwa operations also developed the year-round community of The Depot, near the mouth of the Pukaskwa River. The Depot included family residences, blacksmithy and barns, and doctor's office, as well as staging facilities to receive supplies, food, and horses by steam tug from Michipicoten, and to serve the upriver camps. In winter, mail and supplies were carried by dog team over a trail to White River, a rugged seventy-mile trip. Today, one cabin remains.

At The Depot, pulp logs from the rivers were assembled in great rafts for towing to Abitibi's mills at Sault Ste. Marie. At its height, 400 men lived at The Depot, cutting and driving timber from September to June. The "big tow" was the major means of transporting pulp logs from the Pukaskwa to the Sault and United States mills.

For the big tow, a huge raft of pulpwood was made up from sticks floated down the driving river and at first held in the river mouth by a boom of larger logs, stretched across the river mouth from bank to bank. The ends of the boom were attached to the riverbanks by cable to trees or to bolts drilled and driven into rock. A small tugboat, worked within the bow boom, moving boom ends, snubbing ends to shore, watching for jams and strays. A larger tug that would tow the raft out across the lake soon arrived with the tow booms for the *rafting-out* process.

Booms were long strings of boom logs chained end to end. In early days, boom logs were made locally from large white spruce or white pine; later, Douglas fir and Sitka spruce—

tougher, buoyant, and more resistant to rotting—were shipped from British Columbia. Each boom log was twenty-five feet long, at least thirty inches in diameter. A section was made up of twenty-five boom logs and its chains, or about 650 feet long; these sections could be readily attached to form the larger set.

The tow tug laid out sets of boom to enclose the pulpwood at the mouth of the river, and when the tow boom was secured around the mass of sticks and released from the riverbanks, the rafting-out was complete.

The tow tug pulled the raft out to open water and let out the tow cable—1,400 to 1,800 feet—far enough that the raft would not be pulled against the current generated by the tug's propellers. The speed of the tow: one mile per hour.

Trips across or down the lake were long and—in stormy weather and high seas—arduous and hazardous. At this speed, a headwind often meant that the whole tow moved backward. On one occasion, a tug and its raft were pushed back for a period of twelve days, eventually breaking up on the American south shore.

The size of the rafts varied, but some were huge. Total length of two booms surrounding the raft of up to two miles was not uncommon. The amount of pulpwood in a single raft was commonly around 10,000 cords, in the neighborhood of one million board feet. Some rafts were even larger.

Lost and scattered pulpwood logs commonly were strewn along the shores and islands, generating a small business of "pulp pickers" who beachcombed for the sticks and sold them back to passing tugs. For isolated coastal inhabitants, it was an occupation that supplemented regular income from trapping and fishing.

Numerous pulp sticks may be found today along deserted beaches and coves on the eastern shore, as well as old boom logs with rusting chains still attached—reminders of the "big tows" on Lake Superior.

Logging on the Pukaskwa ceased in 1930. The river drives were over and The Depot closed down, but relics of the old camps and the settlement remain today in Pukaskwa National Park.


With the close of Abitibi's cutting on the Pukaskwa and the eastern shore rivers, and after the depression of the 1930s, the company commenced cutting the last large area included in Francis Clergue's old concession: the vast White River spruce lands, butting up to the Pic River boundary. It was a unique logging operation.

In 1941, Abitibi set up a headquarters depot at Regan, where Canadian Pacific rails crossed White Lake, located on the east shore, near present-day Mobert Indian Reserve. From this headquarters, the company directed and supplied its cutting operations around the north end of White Lake, the northern tributary Shabotik River, and on tributary streams farther down on the White River. Camps were established on White Lake and downstream at several points on the White River, including a rafting camp at the mouth of the river on the shore of Lake Superior. On White Lake, boats of many kinds operated to serve the cutting, bush camps, and log transport—including diesel tugs, amphibian "alligators," scows, and even two LST's from amphibious operations in the Second World War.

The White River drive consisted of assembling pulp logs from the upper lake into rafts on White Lake, using white spruce boom logs, towing by tug on White Lake to Regan (including squeezing these rafts through The Narrows on White Lake). Logs were "spilled" from the rafts at Regan. Then commenced a wild drive from Regan to the river mouth on Superior, something like sixty miles of river, including falls, rapids, and gorges that often created big jams. At the mouth, logs were again assembled into rafts, and the "big tows" were made up again on Lake Superior for the tow to Sault Ste. Marie. Larger rafts, towed by diesel-powered vessels, included up to 14,000 cords of pulp logs.

The peak of the White Lake operations was in the 1950s, and Regan was a busy company town. Abitibi held cutting rights on some 2,000 square miles, and White Lake produced 100,000 to 150,000 cords per year. The last White River drive was in 1964.

Another unique log transportation system was in operation on the Black River from 1938 to 1964. The Black is a major tributary of the Pic River, immediately to the west of the Pukaskwa peninsula. Ontario Paper Company, a subsidiary of the *Chicago Tribune* newspaper, operated a newsprint mill at Therold, Ontario, near Niagara Falls, and acquired a license to cut pulpwood from 781 square miles in the Black River watershed. Since the Black entered the Pic River about two miles upstream from the Pic's mouth, a river drive on the Black would debouch logs into the the lower Pic, potentially mixing with the logs of the Marathon Corporation, which logged the Pic River at that time. So Ontario Paper constructed a flume, three and one-half miles long, to carry its logs from the lower Black to rafting grounds on Lake Superior, separate from Marathon's Pic River logs, on the west side of the Pic. But the Black River emptied into the Pic River on the *east* side, so the flume had to be constructed to *cross over* the Pic, on a bridge 300 feet long and 80 feet above the surface of the Pic River.

The rafting area of Ontario Paper developed into the company town of Heron Bay South. From here, Ontario Paper transported its pulp logs, not by raft but in large lake freighters, to its mills at Therold. The river drive and unique flume operation were gone by 1965.

In the watershed of the Pic River, logging began in the early 1920s. River drives were run from the Pic River camps, and great rafts were towed across Lake Superior to Wisconsin mills. In 1937, Marathon Paper Mills Company, of Wisconsin, was granted cutting rights by the provincial government on 2,500 square miles on the Pic River, with the proviso that the company build a pulp mill in Canada. This it did in 1944, as the Marathon Paper Mills of Canada. Changing ownership to American Can Company and, most recently, to James River Marathon Limited, of Georgia, the mills of Marathon remain as one of the largest paper industries on the North Shore, using spruce and balsam fir logs from the Pic River watershed. The river drives, which once literally filled the lower river, are over; the last on the Pic—one of the last on the North Shore—was in 1983.

Operations on the Little Pic River were conducted in the period 1933-49 by the American-owned Pigeon Timber Company. Six bush camps were in operation, three along the river and three at the river mouth and siding of the Canadian Pacific Railway. In the middle 1940s, the Neys Prisoner of War Camp held German prisoners captured by British forces in the Second World War, with a total of 2,000 prisoners passing through the camp during the war years. The prisoners worked as labor in the bush camps on the Little Pic, each one required to cut his quota of one cord per day. The great bulk of pulpwood cut by Pigeon Timber Company in some years was by prisoner labor. The camp was evacuated in 1946.

After the Marathon mills were in full operation in 1947, logs from the Little Pic drives were rafted a short distance to Marathon, rather than to American mills. The last river drives on the Little Pic were in the late 1950s, and truck hauls carried the Little Pic's logs to Marathon after that. Today, the mouth of the Little Pic is incorporated into Neys Provincial Park.

Near Terrace Bay, the Aguasabon River enters Lake Superior, after dropping through a deep, rocky gash in the ground and driving an electric generating station. Far upstream, the headwaters of the Aguasabon approach Long Lake and the Kenogami River on the other side of the continental divide. The Kenogami flows toward Hudson Bay.

In the 1930s, the Long Lake country contained great quantities of spruce, but for paper-mill developers on Lake Superior, the Kenogami flowed the wrong way. So in 1938, American-owned Longlac Pulp and Paper Company constructed a dam on the Kenogami, raising levels in Long Lake so that water flowed out to the upper Aguasabon—diverting water from the Hudson Bay drainage—and permitting the driving of pulpwood logs to Lake Superior, and to the company's paper mill on the shore.

Today, the pulpwood drives are gone, but the Longlac mill became the large Kimberly Clark Pulp and Paper Company mill at Terrace Bay. And the additional water in the Aguasabon drives the large electric generating station near the river's mouth, started up in 1941. This "Longlac Diversion" adds nearly 2 percent to the outflow of Lake Superior waters through the St. Marys River.

Early logging in the Nipigon basin was primarily for ties and trestle timbers in building the Canadian Pacific Railway in 1883-85. Timber needs for the Canadian Pacific and Canadian Northern railways were also supplied by cutting in the Nipigon area until 1915. And after that, pulpwood logging increased with the development of paper mills in Port Arthur, Sault Ste. Marie, and, more locally, in Nipigon and Red Rock. Driving rivers included the Black Sturgeon and Jackfish, directly to Lake Superior, and the Nipigon River.

The Nipigon Basin was one of the major pulpwood cutting regions. Logs were landed directly on Lake Nipigon or cut farther inland and driven down tributary rivers. Towed in booms across Lake Nipigon to its outlet, logs were driven down the river to Lake Superior and paper mills. Beginning in the 1920s, the Nipigon river drives continued for a half-century, and ended in 1973.

Some of the earliest pulpwood cutting—in the early 1890s, coincident with logging for pine elsewhere—was along Lake Superior shores between the Pic River and Black Bay. The area contained some of the finest spruce on the North Shore. Hazlewood and Whalen Company logged the spruce, and it towed rafts to mills in Port Arthur. Enjoying huge success, Hazlewood and Whalen later flourished as the North Shore Timber Company of Port Arthur Limited, but it did not survive to the present.

Black Bay Peninsula, St. Ignace Island, and associated offshore islands were logged over a long period, from the late 1800s to the early 1970s, mostly for pulpwood but including limited cutting for birch veneer and aspen. Because of transportation difficulties inland, most cutting was confined to areas adjacent to shores and bays. High-quality timber was patchy in these areas, however, and limited access and winter travel to the islands combine to preclude current commercial cutting.

The first paper mill in the Lakehead was that of the Provincial Paper Company Limited, in 1917; others, established by Abitibi and Great Lakes Paper, soon followed. The last-named was destined to become one of the largest pulpwood operations on the North Shore, carried out in the Dog River drainage and associated areas northwest of Thunder Bay.

The Dog River flows from the north through boreal forest that contained immense stands of black spruce. The driving qualities of the Dog River, and its location convenient to the mills of Fort William, were early recognized. Completion of the Canadian Pacific Railway allowed ready transport of logs to Fort William. The original company, Backus Brooks Enterprises of Minneapolis, first acquired 566 square miles in the original Dog River limits; later, as Great Lakes Forest Products of Thunder Bay, the company added much more.

The Dog River flows through Dog Lake and later empties into the Kaministiquia River, which flows to Thunder Bay. However, the river drives came only partway down, to above Dog Lake, where pulp logs were removed and transported overland to rail landings. The lower reaches of the Dog River were sluggish and not considered good driving water.

In the late 1940s, a series of logging camps were established along the river, serviced by trucks that could travel a winter tote road parallel to the river. But the subsequent Dog River Road, completed in 1959 and later upgraded for all-weather travel by semi-trailers in 1966, allowed year-round transport of logs from camps to rails. So the last river drive was completed in 1960.

Eventually Great Lakes Forest Products held licenses for about 15,000 square miles of timber. The mills, now diversified to produce much more than paper, use more than one million cords of wood annually.

The change in vegetative cover along the North Shore had been drastic. For a fifty-year period, human industry had concentrated first on quick removal of the tall pines and, second, on long-term, professional management of spruce-fir forests for permanent paper and wood-products industries. Timber removal, fire suppression, modern transportation, and human settlement all combined to bring about a new environment along the North Shore of Lake Superior.

White pine has not returned. The European blister rust has prevented both natural regeneration and cultivated stands, although modern attempts to develop resistant strains of white pine may soon be successful.

Red pine does well when planted, and many plantations have been established, but the species can no longer naturally reseed itself on a large scale.

So the new forest mosaic north of Superior is a different mixture now, with a shift toward deciduous species and smaller conifers.

Wildlife populations, too, changed significantly—both in abundance and species distribution—in accordance with different forest habitats.

Commercial resource exploitation had turned from the land to the water, to the fish resource of the great inland sea—again, seemingly, inexhaustible.

Most important, however, were changes in human perspectives and attitudes toward North Shore resources. The sobering experience of the pine timber removal, the development of conservation attitudes, the recognition of recreational values by new visitors, and the growing scientific effort directed at every type of resource—all created a new atmosphere of stewardship.

# EIGHT

# Fish Story I—The Natives

When the first adventuring Europeans entered the North Shore region of Lake Superior in the early to middle 1600s they found the Ojibway heavily dependent upon fish for food. The abundance of fish taken impressed the newcomers greatly. Their accounts, though often exaggerated, bespoke a rich productivity of this aquatic resource. Legendary are the relations of Indian dip netting for whitefish in the rapids of the St. Marys River and, as well, the frenzied activities in Indian summer fishing villages when lake trout migrated into shallow waters.

Explorers, traders, and, eventually, early settlers followed suit. Louis Agassiz, famed geologist and naturalist, in making his classic scientific tour of the North Shore in 1848, also remarked upon the great abundance of fish. Fish continued to be a staple item in human sustenance on the North Shore. The supply appeared inexhaustible, the lake vast, almost infinite.

To compare these early accounts to the current fishery along the North Shore—or of Lake Superior and, indeed, the Great Lakes—evokes the question: What happened?

Unraveling that question reveals a litany of inertia, wishful thinking, and biological tragedy. Certainly, it is an account—told only briefly here—of one of the most damaging of human impacts upon a natural resource. Yet it is also a story of daring and dedication, scholarship and research, and at least modest success toward its solution.

The story, however, remains unfinished, the fish population exceedingly dynamic and changing.

Early reports of the abundance of fish in Lake Superior are misleading. Indian fishing history, reports of early travelers, and accounts and records of fur traders were all effusive in their relation of great quantities and sizes of fish taken, mostly whitefish and lake trout. However, Lake Superior is an oligotrophic lake—low in nutrients and food productivity. How could such infertile waters produce such immense quantities of fish?

The answer lies partly in the fact that the fishing was done at times and places of *concentration*, when the fish were highly vulnerable—in spawning runs and during seasonal migrations—not at random over the entire lake. This congregation of fish at times of spawning migrations gave an inflated image of the lake's productivity. For these great fish concentrations meant that a very much larger area of lake must have been involved in producing the fish and their food organisms. Lake Superior is a biologically "poor" lake and cannot possibly produce throughout its vastness the fish abundance observed near shore and in tributary rivers by early inhabitants and travelers.

Early Americans, limited to inshore fishing, had long ago discovered the most efficient places and times for harvesting whitefish and lake trout, and explorers and traders emulated the Ojibway.

The more intense exploitation of these fish concentrations by white commercial fishing resulted in drastic reductions in certain fish populations, even though such efforts seemed to be so small compared with the total area of Lake Superior. The apparently highly productive fish community was instead a fragile resource.

Furthermore, when fish populations remain in essentially an unexploited state for a long time—as had been the case in Lake

*Gargantua Harbour*

Superior—large, long-lived individuals accumulate. Then when an intense fishery commences, such as the white man's exploitation, the large fish are often removed preferentially. In the face of continuing intense fishing, these older fish in Lake Superior no longer accumulated.

Ichthyologists and fish biologists now conclude that Lake Superior and its tributary waters contained at least seventy-one native species of fish. As best we know, in presettlement times Lake Superior alone held thirty-two of these native species; seven more have been introduced intentionally and five more accidentally, for a total of forty-four today. In addition, thirty-nine native species inhabited tributary streams, but not Lake Superior. Thus the Lake Superior Basin today contains a total of eighty-three species.

Even so, this seemingly large number does not accurately reflect the importance of Lake Superior's fish fauna. For one thing, even though Superior is the largest of the Great Lakes, the number of fish species is small compared with that in the other lakes—

about half that in any one of the other lakes, only one-fourth of the total native species in all Great Lakes (177 species). On the other hand, the existence of forty-four species in Lake Superior still suggests a complex basis for the major elements in the succession of fish populations on the North Shore—a succession more complex than we might think.

Changes that occurred hinged chiefly on two species: the lake whitefish and the lake trout. Significant changes in stocks of most other species depended, either directly or indirectly, upon basic changes in whitefish and trout populations or in fishing efforts applied to them.

Prehistoric dependence upon fish along the North Shore also largely involved these two species. Trout congregated in great numbers and large sizes during spawning runs and migrations to shallow inshore waters, river estuaries, or connecting streams; whitefish, more restricted to shallow waters year-round, also depended on foods and other environmental factors near shore.

In shallow waters, both whitefish and trout were vulnerable to the natives' primitive fishing technology and limited accessibility by birchbark canoe. Fish of both species could be harvested at a time of year (late summer to late fall) when fish could be preserved, especially by freezing. And, consciously or not, the Ojibway exerted a degree of selectivity toward fish with high oil content—and which were thus more nutritious and energy-rich.

Some of the original native species remain in Lake Superior today at much reduced levels. However, little note is taken by human exploiters of the smaller or rare species, of the fact that many species inhabit only tributary streams and a few marshy shorelines, or that some major fish groups—for example, the sunfishes and minnows—are esentially unrepresented in Lake Superior's waters. Others, like the small but abundant bottom-dwelling sculpin and several chubs, were always important as essential prey for the predaceous lake trout but were unknown or little recognized as important in an earlier era.

The lake trout—*Salvelinus namaycush*—is mainly a denizen of nearshore waters. Of all our native salmonids, however, the lake trout's preferred temperature is the lowest, and it inhabits the cold water of Lake Superior's depths in midsummer. Viewed from above, the lake trout is long and lean—not unlike the northern pike—dark gray-green on the back with the light-colored vermiculations (wormlike marks) that are characteristic of its genus. Inside, its flesh is often pink to orange-pink, or ranging up to bright red-orange.

In late summer and fall, the lake trout migrates into shallower waters to spawn over rocky shoals and reefs. A few populations on the Canadian shore, however, are anadromous—lake living but stream spawning—migrating upstream for several miles to spawn.

The lake trout is not the simple population unit that might be supposed. Two subspecies are recognized, plus other distinct stocks that are nearly isolated reproductively. Many quasi-discrete stocks are separately recognized by appearance in shape and color, place of spawning, or time of migrations. The river-running lake trout occur in separate stocks distinct from major lake populations; the river stocks are distinct from each other as well, making spawning runs at different times.

The lake trout is a fish predator—or piscivore—and depends heavily upon the abundance and availability of smaller prey fish at the proper time and depth.

As human food, and more recently for sport, the lake trout was a major fish species in Lake Superior; it remains today the principal species of interest, though far from many other species in abundance.

Lake trout occur in four major groups, distinctively recognized. By far the most abundant is the *lean* lake trout, the common, regular lake trout that we see or (if we're lucky) sometimes eat; *leans* are in the subspecies *Salvelinus namaycush namaycush*. In contrast, a second subspecies, much less common, is the deepwater, heavy-bodied, much oilier *fat*, or siscowet, subspecies *Salvelinus namaycush siscowet*. The siscowet is so oily—50 percent of its body weight is in fat—that it is practically unpalatable and was used only for salting and smoking. The Ojibway name *siscowet* means "cooks itself," in reference to its high oil content.

A third type, known as *half-breed*, was distinguished by fishermen as an intergrade between *leans* and *fats*, but it is fairly well identified now by biologists as a small, immature *fat*. The fourth, known as a *humper* in the United States or *paper-belly* in Canada, is apparently a *lean* that spawns over isolated offshore reefs and shoals and has become partially isolated reproductively; it has a white, paper-thin abdominal skin.

The *lean*, however, is the regular object of commercial fishing or individual sport and food use, and the only one that most of us will ever know.

Even so, within the *leans* there appeared many quasi-discrete stocks with different spawning times and locations. Early commercial fishermen recognized these stocks as "breeds," identifiable by color of fins, skin, and flesh, by shape and size, and by spawning time and location. As many as thirty-six different breeds were recognized, with such colorful and quaint names as "redfins" and "yellowfins," "moss trout" and "sand trout," "blacks," and "racers."

Two hundred distinct spawning grounds have been located, including twenty rivers, where lake trout "home" to their natal reefs, shoals, or river rapids. Spawning grounds with surviving stocks include those around Isle Royale, the Slate Islands, and Pie Island. And although the river-spawning stocks were depleted before 1950, Canada lists eight rivers still being used for spawning.

The lake whitefish—*Coregonus clupeaformis*—is a much different fish. It is restricted to a shallow inshore habitat year-round. It is smaller, and it is not a piscivore, feeding instead on small invertebrate animals that live at the bottom, and consequently it is not taken by hook and line as easily as the lake trout. It is not separated into subspecies or recognizable variants, but as with the trout some quasi-discrete populations occur, including some river-running stocks.

The whitefish is also much different in its appearance—being totally white, even fulgent when first removed from the water. Its flesh, too, is white. When prepared really fresh, the whitefish is one of the most delicious of all freshwater fishes. Its mouth is somewhat inferior—that is with an overhanging snout—an adaptation to its bottom-feeding bahavior.

The whitefish also spawns in the fall and is vulnerable at that time to netting in shallow water and connecting streams. In early fishing days, the lake whitefish was by far the most numerous food fish taken. Declines in whitefish populations occurring in the late 1800s and early 1900s may have been partly due to the suffocation of eggs by woody waste, such as sawdust and bark, from lumbering operations in that period.

Other species closely related to and resembling the whitefish in their silvery coloration have been important in Lake Superior. These included the lake herring, or shallow-water cisco— *Coregonus artedii*—which is somewhat smaller than the whitefish. The herring is a pelagic fish, coursing in open water far from shore but not far below the surface; thus it is known as a "shallow-water cisco."

In its pelagic existence, the herring is a plankton feeder and thus a major link in the food conversion chain. It occurred in enormous abundance and provided a major food resource for the lake trout. The herring was exploited by commercial fisheries after declines occurred in whitefish populations, and for several decades prior to the 1950s it was taken in numbers many times higher than even those for the whitefish. It also is a fall spawner; at this time, the herring may be taken by the modern angler with flies and is reported to be a strong fighter on the end of a line.

Several species of small whitefishlike chubs were present as native fishes in Lake Superior, in the same genus as whitefish and herring—*Coregonus*. Also living far from shore, the chubs, unlike the herring, are more demersal—associated with the bottom— and thus some are known as deepwater ciscoes.

The chubs originally comprised five species in Lake Superior, of which only three occurred in high numbers. The five chubs were similar and not always distinguished by fishermen, and, indeed, they are genetically flexible, with hybridization occurring commonly. The bloater—*Coregonus hoyi*—was the smallest of the chubs, seldom more than eight inches long, and is essentially the only species of chub left.

All the above species are high in oil content and were considered the choice fishes. In prehistoric times, the largest—lake trout and whitefish—were the most valuable to the natives: energy-efficient with respect to their large size and ready availability in inshore waters, and energy-rich in their high oil content.

Whitefish and lake trout were much more abundant on the eastern and northern portions of the North Shore, with their more irregular shorelines and more shoals, than on Minnesota's coast with its straighter shore and relative lack of shallow areas. Lake herring, however, were for many years more abundant in Minnesota waters.

Other native fishes were common and important as human food but, with lean flesh and lower oil content, were never as important in Indian fisheries. The lake sturgeon (*Acipenser fulvescens*) is a large fish that was taken by the natives. When fisheries were developed by the whites in the latter 1800s, sturgeon were intensively exploited for their roe—and also for gelatin from the inner lining of the gas bladder, used in making isinglass windows. The sturgeon is a spring spawner and moved then into streams, including the rapids in the St. Marys River, where it was easily taken. The sturgeon is a long-lived fish—100 years is not uncommon—and was thus subject to rapid declines under intense fishing pressure. Fouling nets set for other species, the sturgeon was considered a nuisance, and it was deliberately overharvested in order to eliminate it. The sturgeon, too, may have been subjected to deleterious effects of lumbering wastes. The species was virtually gone from Lake Superior by 1940.

Both the northern pike, or jackfish (*Esox lucius*), and the walleye, or yellow pickerel (*Stizostedion vitreum*), are native in Lake Superior and along the North Shore. But they are species much more adapted to shallow, calm shores and, especially, inland lakes and river estuaries, rather than open waters of the big lake. Both are large piscivores and are thus subject to hook-and-line fishing. Except locally, they have not been numerically important in either the sport or commercial fisheries of Lake Superior. Their flesh is lean and thus not as important in native fisheries as the oilier trout and whitefish.

The burbot, or freshwater cod (*Lota lota*), occupied cold, deepwater areas of the lake, as did lake trout, but was less abundant. Like the lake trout, the burbot is a piscivore, preying on the same food fishes in deep water. Little favored as food, it was lightly fished.

The highly regarded brook, or speckled, trout (*Salvelinus fontinalis*) is also an original inhabitant of Lake Superior. As the genus name suggests, it is a close relative of the lake trout. The brookie, however, is primarily a stream fish, not found commonly in the lake's open waters. Originally, it occupied only those portions of streams below natural barriers and waterfalls, having apparently arrived in Lake Superior after the drop in glacial waters to levels below barriers. It has since been introduced to most upper reaches of tributary streams, where it is the principal target, even though small, of bush-whacking stream anglers.

Two species of suckers have apparently been abundant but unutilized components of lake populations. The longnose sucker (*Catostomus catostomus*) and white sucker (*Catostomus commersoni*) are currently under study to determine whether they might be more efficiently harvested and used as food in various fish products.

All of the foregoing is not to imply that no nonnative fish species occurred prior to 1950. To the contrary, highly prized fishes like the brown trout and rainbow trout were successfully introduced as early as the late 1800s. The rainbow smelt, first introduced in the Great Lakes in the early 1900s, later became important in both commercial and recreational fishing in Lake Superior. However, major changes due to nonnative introductions did not occur until later.

We must now interupt this account of fish along the North Shore, as, indeed, were the fish populations themselves disrupted. No longer were the vast quantities of lake trout and whitefish—which since prehistoric time successively served the Indians, fur traders, and resident white families—to continue as viable resources. No more would these delectable and nutritious food fishes nurture human exploiters or the fishing industry that once dominated the economics of the North Shore, as they had in centuries past.

Through artificial changes—in old water routes, passages cut through ancient geological barriers, came an invader from the Atlantic Ocean—unseen at first, insidious, predatory, eventually overwhelming. The invader was to change the fish resource of the North Shore, and of the entire upper Great Lakes, drastically and perhaps irrevocably.

# NINE

# *Invader from the Sea*

On a November day in 1824, the president of the Welland Canal Company turned a ceremonial spadeful of earth, officially beginning the construction of the Welland Canal. This canal would cross the Niagara Peninsula to circumvent Niagara Falls, connect Lake Ontario to Lake Erie, and open watercraft navigation from the ocean to the upper Great Lakes. The assembled crowd gave a rousing three cheers.

Exactly five years later, in November 1829, two sailing schooners, one Canadian and one American, were towed from Lake Ontario past cheering crowds, up through thirty-four locks, and, via the Welland and Niagara rivers, into Lake Erie. At Buffalo, at the eastern end of Lake Erie, five days after entering the locks, the two schooners arrived to more cheers from citizens gathered on the wharves, and to a roaring artillery salute as well. Navigation, from salty ocean to the freshwater seas of midcontinent, was a reality.

Improvements were made over the ensuing years. Primitive locks were upgraded, further enlarged, and deepened. For a second, third, and fourth time, the canal was reconstructed—in 1853, 1887, 1933. Thus, a little more than 100 years after completion of the first canal, the fourth was ready to serve as the model for the St. Lawrence Seaway, scheduled for completion in

1959. This important link in the Seaway—now the Welland Ship Canal—opened much of the Great Lakes to ocean-going shipping from around the world. Today's canal comprises eight locks, raising ship traffic 326 feet along a linear distance of 28 miles.

But along with the barges, boats, schooners, and steamships—some time in this hundred-year period, the exact date unknown—came another saltwater traveler. Slipping through the locks unseen, perhaps at night, perhaps in the shadow of a ship, or—most probably—attached as a hitchhiker to the ship's hull, this marine traveler made its way from Lake Ontario to the upper Great Lakes. It adapted with remarkable success.

This was the sea lamprey—*Petromyzon marinus*—a predatory snakelike fish that was inevitably to become a menace, dreaded alike by commercial fishermen, resort owners, and sports anglers along the North Shore—and throughout the Great Lakes.

Upon the aquatic life of the Great Lakes, the introduction of the sea lamprey was to have the most profound ecological effect in the history of the lakes. The intricate interrelationship among the populations of lake trout and whitefish and associated species that had existed for several millennia—serving prehistoric Indians, explorers, and early white settlers along the North Shore—was about to be shattered.

There were no cheers for the sea lamprey.

Already the lamprey was no stranger to fresh water. Well established in Lake Champlain and the Hudson River drainage, in apparent ecological balance with other fishes including the lake trout, the sea lamprey has existed for a long time, possibly since early postglacial times. The first artificial connection between the Atlantic Ocean and the Great Lakes—the Erie Canal, completed in 1823—perhaps provided the first route for incursion of the sea lamprey into Lake Ontario. In 1835, the first specimen was observed in a tributary of this lowermost Great Lake.

The St. Lawrence River, of course, was a more direct route, but the lamprey apparently was never able to navigate earlier existing falls and rapids in the river.

Between Lakes Ontario and Erie, the massive Niagara es-

*Sand River*

carpment and great falls presented an insuperable barrier to human navigation—and apparently to the sea lamprey as well—until the Welland Canal provided an easy passage to both. Thereafter, the invasion of the lamprey into the other four Great Lakes was one of rapid dispersal.

First observed above Niagara Falls in Lake Erie in 1921, the lamprey was successively reported in Lake Michigan in 1936 and Lake Huron in 1937.

The first confirmed report in Lake Superior was from near Isle Royale in August 1946. In 1949, there was another report from Ontario's Montreal River, and in the same year Minnesota commercial fishermen reported lampreys from the western end of Lake Superior. It seems probable that the lamprey, unseen,

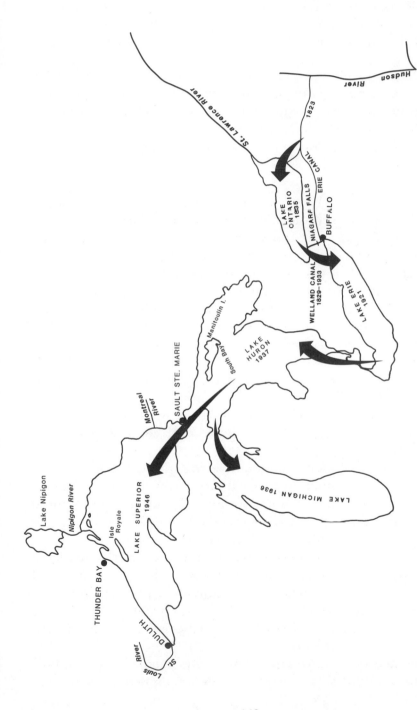

Route taken by the sea lamprey into the Great Lakes. This primitive, saltwater fish from the Atlantic Ocean apparently entered the lakes by way of the Hudson River and the Erie Canal to Lake Ontario, and to the upper lakes by way of the Welland Canal around Niagara Falls. (Redrawn from Downs 1982, with permission of the University of Wisconsin Sea Grant Institute.)

had colonized the upper lakes extremely rapidly, and by the late 1940s it was well enough established, especially in Lakes Huron and Michigan, to have already caused damage to fish populations.

Thus it was that by 1950 there came the full realization that in the making was a natural resource disaster of alarming proportions. Immediate efforts were initiated by the fisheries profession in both the United States and Canada for control of the sea lamprey. It would prove to be a monumental task.

The sea lamprey is a primitive fish of ancient lineage, probably several million years. In a family of jawless fishes, the lamprey has instead a suckerlike mouth; inside are rows of rasping teeth. In the adult stage, the lamprey attaches to its prey fish with the sucker mouth, rasps an opening, and extracts blood and other body fluids. It is to be classified as a predator, rather than a parasite, for an attack usually results in death of the prey, and the lamprey moves on to another victim. The lamprey is not an *eel* (which has jaws, like our other better-known bony fishes), and it is not properly termed a "lamprey eel."

Other, native species of lampreys occur in Lake Superior and its northern tributaries, but these are smaller and either are nonpredaceous (two species of brook lampreys that feed on algae and detritus throughout life) or have evolved in balance with their prey so that they coexist.

In spring, as water temperatures of tributary streams rise, adult sea lampreys migrate upstream to spawn. From the Atlantic Ocean's salt water to freshwater streams, this movement required a distinct physiological change, for osmotic pressures are reversed between salt and fresh water. But when this primitive fish dispersed to freshwater lakes, the adaptation to a life-style totally in fresh water apparently was easily effected. In the stream, a nest of pebbles is constructed by carrying individual pebbles and small stones in the lamprey's sucker mouth. Eggs are deposited by the female in the nest, fertilized by the male, and, as with Pacific salmon, the adults then die.

Hatched one to two weeks later, tiny, needlelike larvae burrow into sandy or silty substrate in the streambed, and there as juvenile lampreys—ammocoetes—they undergo a long period of development and transformation.

Ammocoetes are neither parasitic nor predaceous during this time, feeding on organic detritus and plant material. Larval development is slow; depending on water temperature and food availability, a cohort of ammocoetes may require from three to eighteen years, a factor that—as we will presently see—provides a distinct advantage in lamprey control operations.

Upon final larval development, emergence from the silty streambed takes place, and the newly transformed lamprey moves downstream to the lake or ocean, usually in winter. In open-water depths, the sea lamprey becomes a free-ranging predator, attacking its prey to grow further into a fully developed, sexually mature adult. Its final size: fifteen to twenty inches long. The predaceous stage typically lasts about a year and a half.

We know very little about where the adult spends its hunting period, and how far and deep it roams in Lake Superior. We see wounds and scars on fishes whose habits we know better, and we can therefore make some inferences. For example, we know the lamprey travels primarily fairly close to shore; very few have been collected in the open lake.

And in some early spring, five to twenty years after its own birth, the adult repeats the upstream migration of its own parents and spawns to complete the life cycle.

The immediate effect of the lamprey invasion of the upper Great Lakes was devastating. Precipitous declines in lake trout populations and other fishes proceeded up the Great Lakes in lock-step with the lamprey's dispersal and increase.

A major factor was the lamprey's predilection for attacking larger fish, rarely preying on lake trout less than seventeen inches long, in Lake Superior about four years old. Because up to 90 percent of actual lamprey attacks are fatal, that portion of the trout population older than four years was decimated. And since sexual maturity and spawning in the trout are first attained only among older and larger members six to seven years old, reproduction was brought to a virtual halt.

As a newcomer in an environment in which it had not evolved, preying on the only two species of predator fishes—lake

151

trout and burbot—the sea lamprey faced no natural control. Its fecundity was high—30,000 to 50,000 eggs per female. Many suitable spawning streams were available. Reliable estimates suggest that during the 1940s and 1950s, the period when its populations in Lake Superior were rapidly developing, the number of lampreys increased twenty-to fiftyfold per generation.

By the late 1940s, lake trout in Lakes Huron and Michigan were essentially extirpated. In Huron, the first of the upper lakes to be affected, commercial harvest of lake trout plummeted to 1 to 2 percent of former levels. Since the lamprey invasion into the Great Lakes was progressive from east to west, Lake Superior was the last to be affected; finally, in the years 1953-62, trout populations in the uppermost lake fell more than 25 percent per year. Although a viable population of native trout remained in Superior, numbers were only at a small fraction of former abundance. The commercial fishery, exploiting mainly larger fish, collapsed, and many fishing industries of long standing on the North Shore ceased operations.

During the decade of the 1950s, the lake trout commercial catch dropped to only about 10 percent of former levels, and in 1962, except for permit fishing to obtain assessment data, commercial fishing was prohibited.

The lamprey attacked fishes other than lake trout, too. Starting with the largest species, the lamprey preyed upon the burbot or freshwater cod, the lake whitefish, suckers, lake herring, and even large chubs.

In the early 1950s, when the potential disaster was beginning to be realized, some controversy developed as to whether it was really the lamprey that was responsible, or whether the lake trout had been overexploited by commercial fishing. Strong circumstantial evidence, critically examined, brought a definite indictment against the lamprey. For one thing, the progressive decline of trout through the Great Lakes coincided precisely with the increase of lamprey, while fishing had been in operation for a much longer time. The burbot, a large predator that occurred in great numbers but had not been fished for, declined in abundance in parallel with the trout. In South Bay, on Manitoulin Island in Lake Huron, commercial fishing had been closed, but here, too, the lake trout was exterminated as the lamprey dispersed through the region.

The finger of guilt appeared firmly pointed at *Petromyzon marinus.*

Once the full impact of the disaster was absorbed, and the basic outline of lamprey life history understood, it seemed that the most promising point of attack for lamprey control was at the time of spawning, when migrating adults concentrated into suitable streams.

The number of rivers and streams suitable for successful spawning is limited. Rocky, torrential streams are not suitable, because the lamprey, unlike trout and salmon, cannot leap waterfalls. Furthermore, the sand-gravel substrates necessary for ammocoete development are not sufficiently present in high-gradient streams. Consequently, few streams on the western part of the North Shore, in the Minnesota portion, with their tumbling, torrential character, were suitable; many more along the eastern side of the lake in Ontario were high lamprey producers. Contributing to Lake Superior as a whole, streams on the south shore, in Michigan and Wisconsin, were also heavy producers. Control efforts immediately became an international effort.

Joint efforts of Canada and the United States as early as 1946 resulted in the international Great Lakes Sea Lamprey Committee. Recognizing problems broader than the lamprey, this body evolved to the Great Lakes Fishery Committee in 1953 and then, by treaty between the two countries in 1954, to the establishment of the Great Lakes Fishery Commission in 1955. The commission functions to advise both governments on all aspects of Great Lakes fisheries; its agents—the United States Fish and Wildlife Service and Canada's Department of Fisheries and Oceans—are charged specifically with the responsibility for sea lamprey research and control.

At first, control efforts were directed toward eliminating the spawning run of adults, concentrated as they were in lower stream reaches. Mechanical weirs and traps, dams, and electrical killing devices were employed to remove hundreds of thousands of adult sea lampreys, all ready to spawn. But despite this great effort, the adult-removal program was unsuccessful. Problems of constant maintenance, access to remote areas on the North Shore,

153

floods that destroyed dams and weirs—all combined to prevent sufficient and reliable control.

Meanwhile, another sensitive point in the lamprey's life history was under study, namely the time when high concentrations of ammocoetes could be found in spawning streams. Here the long larval life of several years might be an advantage, for if a means such as a selective toxicant, killing lamprey larvae but not other fish, could be found and employed, many generations of lampreys could be destroyed with one treatment, without harm to fish of other species.

At a United States Fish and Wildlife Service research facility, the Hammond Bay Biological Station in Michigan, the service's scientist Dr. Vernon C. Applegate in the early 1950s searched for a solution. Applegate and his team screened 6,000 chemical compounds—some common, like an extract of black walnut hulls; some exotic, like complex synthetic organics—for their relative toxicity to lamprey ammocoetes and other fishes. The number of compounds tested later rose to 10,000.

A group of related compounds was found to do the job. It was selective in its action because, while potentially toxic to all fishes including lampreys, the material was detoxified by the livers of bony fishes like trout and sunfishes, but not by the more primitive lamprey.

The most promising member of this chemical group carried the jaw-breaking scientific name 3-trifluoromethyl-4-nitrophenol. To save the jaws of reporters for decades to follow, an appellation—TFM—was quickly adopted. TFM was expensive, and later a cheap synergist, actually a molluscide known as Bayer 73, was co-applied at time of treatment to increase the efficacy of TFM and reduce costs.

Many laboratory bioassays and pilot field studies were conducted in the ensuing few years, and results were promising. The question of where to start extensive control operations in the Great Lakes was soon decided: with lake trout essentially gone from Lakes Huron and Michigan, the decision to attempt to save the remaining viable population of trout in Lake Superior was obvious. Under authority of the Great Lakes Fishery Commission, the chemical control of the sea lamprey began in 1958. There were hopes, though remote, for eradication.

Application procedures were not simple. Concentrations of

TFM required under actual field conditions vary—with stream flow, floods, turbidity, water temperature, and other chemical conditions. Consequently very complex procedures are employed in stream treatment operations, involving large crews, mobile laboratories, and continuous operation of test equipment. Supplementary application points are established to ensure proper TFM concentrations at all times during treatment. And, finally, evaluation of the degree of ammocoete kill must be made.

The program began with field treatment of twelve Lake Superior tributaries, and in the next two years sixty more were treated. Many treatments were repeated.

The immediate results were striking. Some of the earlier traps and weirs continued to be used, not for purposes of control now, but rather to evaluate the chemical treatment. By 1962 the number of test-trapped adult lampreys in Lake Superior was down to only 13 percent of that in the previous year, the peak. By 1966, the lamprey catch had dropped to only 5 percent.

Large-scale treatments were extended to Lakes Huron and Michigan. Attainment of the goal of successful lamprey control appeared well under way.

On the North Shore, forty-four streams were identified as lamprey producers and treated with TFM. Only four of these—the Gooseberry, Split Rock, Poplar, and Brule rivers—were in Minnesota; in these four, lamprey control appeared to be largely successful. Of the other forty in Ontario, fifteen experienced significant lamprey decreases, and, of these, ten are considered to have been eradicated of sea lampreys. Among those eradicated are the Sand, Dog, Little Pic, Pays Plat, McIntyre, and Neebing rivers. Others for various reasons remain problems and continued targets of treatment: these include the well-known Sable, Pancake, Michipicoten, Steel, Big Gravel, Cypress, Jackfish, and Nipigon rivers, as well as the Pigeon, the border river.

Many more streams have been treated on the south shore, in Michigan and Wisconsin, in which control operations have been similarly successful.

Much concern arose over the possible effects that TFM treatments might have on other fishes. While the original screening of chemicals at Hammond Bay showed a high degree of selectivity toward the lamprey, TFM concentrations not too much higher than those required to kill lampreys will also kill other fishes.

Since all fishes are most exposed to the control operation in their spawning aggregations, TFM treatments are timed as far as possible to avoid spawning times of other species. So far, there has been no effect on total populations of other fish species.

Similarly, some concern has been expressed about possible effects on stream invertebrates, like aquatic insects, that serve as food for fish. Some major invertebrate kills have occurred, but no obvious long-term effects have been demonstrated. Research is under way to further assess this question, particularly with respect to some mayflies that are of special interest.

The difference in physiological response between the sea lamprey and bony fishes does not extend between the sea lamprey and other lampreys. Other lamprey species can be affected. However, since habitat differences exist among these several lamprey species, no major problems have been encountered. For example, brook lampreys mainly occupy stream headwaters, and since the sea lamprey cannot ascend barrier falls, and treatments are limited to lower reaches, major populations of brook lampreys are little affected by treatments.

The next step was the attempt to rehabilitate lake trout stocks by supplementing natural reproduction with hatchery-reared fish. New hatcheries were constructed. Eggs were taken from females of native trout, fertilized, and reared with artificial foods, in a massive effort to restore trout stocks to former levels. By the mid-1980s, more than sixty million young lake trout had been planted in Lake Superior, including about twenty-five million in North Shore waters.

At first, these plantings were not as successful as hoped. Survival was poor at the start, because the fish planted were too young. Reproduction from the planted fish, when they did reach maturity six years after stocking, was not as good as expected; the reason—we've now learned—is that lake trout *home*, much like salmon, and early plantings were not always made near suitable spawning reefs.

Neither was reproduction very good from the surviving populations of native trout. We are not sure exactly why, but changed and new relationships between the trout and other fish species, native and introduced, are suspected.

Intensive research into many factors, under way since the lamprey problem was recognized, is gradually producing solutions. For example, we have learned more about different genetic stocks from different parts of the lake and which stocks to use in rehabilitation.

Successful as the lamprey control program has been, however, it is not unqualified. Some important problems and concerns remain. For example, some large and complex river systems, such as the Nipigon and St. Louis rivers, are extremely difficult to treat. Furthermore, some reports indicate ammocoete production in Lake Superior out from stream mouths—areas that would be almost impossible to treat.

The sea lamprey is especially sensitive to pollution, demanding clean water. Ironically, when some pollution problems were cleaned up in the past several decades as a result of efforts by both Canada and the United States, some streams that previously were relatively free of sea lampreys began to produce the ammocoetes in great numbers. Again, the St. Louis and Nipigon were foremost in this regard.

In view of the remote likelihood of complete lamprey eradication in the upper Great Lakes, the control program must proceed indefinitely, with continually rising costs.

Fortunately, no TFM-resistant strain of sea lamprey has developed, an early fear.

Finally, concern is increasingly expressed about the release of any exotic chemical into our environment, especially into as large and valuable a resource as Lake Superior. Caution has continually been the watchword, and extensive assessment of possible detrimental effects has been conducted. So far, no serious problems have been uncovered. But caution and ongoing research must continue.

Alternatives to chemical control have been proposed. For example, trapping and killing weirs might still be used in some streams to remove spawning adults. And since sea lampreys cannot surmount even small waterfalls, barrier dams, properly designed, could in some streams prevent adult spawning runs and still permit runs of other fishes. Other proposals, still on biological drawing boards, include the sterilization and release of

157

male lampreys; genetic engineering to produce larvae that fail to mature; use of sex attractants, or pheromones, to disrupt lamprey spawning cycles; and the development of lamprey-resistant strains of lake trout. However, these proposals remain speculative.

Although we know little of adult sea lamprey movements in Lake Superior, it appears that mainly they travel in nearshore waters. Some stocks of lake trout, however, appear to spawn and live out their life cycle in offshore waters near distant reefs and islands; these stocks are being investigated further with the hope that, through encouragement of such physical separation, some lake trout stocks may be kept relatively immune to lamprey attack. The red-fin strain, in Ontario waters, is an example.

The problem is not simple, and the long-range solution cannot be simplistic. It is no longer an isolated problem of sea lamprey versus lake trout, as we will see in the next chapter. Increasingly accepted is the view of the Great Lakes, including Lake Superior and its North Shore, as a single *ecosystem*—the holistic approach—with all problems included and many integrated solutions, a complex but encouraging plan.


With all its problems and future uncertainties, the sea lamprey control and trout and whitefish rehabilitation program in the Great Lakes is recognized today as a major achievement in natural resource management. Leading scientists have termed it "one of the great success stories in modern applied ecology."

Costs have been great—25 million dollars per year. But attributable benefits to the Lakes' fisheries have been much greater: 500 million dollars per year—a benefit-cost ratio of 20:1.

Lamprey populations are down to only 5 to 10 percent of peak years; mortality of lake trout is down dramatically, and naturally reproduced trout now make up nearly half the catch. Other species have similarly responded. The program, under the Great Lakes Fishery Commission, is a model of international cooperation.

The fishery of the North Shore is a new fishery—different now with new species, dynamic and still changing, its precise future still uncertain, but still high in potential productivity. A vast and increasing army of scientists, resource managers, and administrators, from universities, federal, provincial, state, and international agencies, continue the fight. Basic research, laboratory experimentation, field operations—all combine in the attempt to preserve and enhance one of the continent's most valuable natural resources.

# TEN

# *Fish Story II—Survivors and Strangers*

Fish populations in Lake Superior and along the North Shore underwent many drastic changes in the middle of the twentieth century. This is not to say that all species and their interactions had been stable and unchanging until then—far from it. From the time of first white settlement to 1950, a period of about 100 years, human activities such as logging and lumbering, planting new fish species, pollution, and, especially, commercial fishing—all combined to produce gradual and generally unnoticed changes. A new sort of evolution had taken place.

But it is now generally accepted that, by 1950, the fish community of Lake Superior had been brought to the sharp edge of a catastrophe, ready to give way at any time. A system of instability, lacking the checks and balances that had evolved over millennia of natural evolution, had gradually crept up over a mere 100 years, unseen in the euphoria of abundance.

The sea lamprey was the factor that caused the sharp edge to crumble, and the biological system collapsed.

Of principal concern and interest in post-lamprey fish populations along the North Shore has been the lake

*Pink salmon fishing, Temperance River*

trout. Most expense and effort in fish rehabilitation has been toward this favored species. The lake trout was a prominent figure in pre-lamprey native populations, and it became the most important species in both commercial and sport fisheries. Many argue that it is the most delectable. After three decades of intensive effort, the struggle to return the lake trout to its previous eminent position now appears to be paying off.

The decision to concentrate sea lamprey control efforts on Lake Superior, in order to save the remaining stocks of lake trout in the Great Lakes, turned out to be a wise one. The genetic pools of native stocks were thus preserved. By gaining time in which to learn more about spawning habits and locations of quasi-discrete populations, researchers have been able to fine-tune fish culture practices and procedures in planting hatchery-reared trout.

There is abundant evidence that the native lake trout today

is making a comeback. Naturally reproduced trout make up nearly one-half of the commercial and sport catch, a great improvement over just a few years ago, when hatchery-produced trout by far dominated the catch.

Yet the lake trout remains a fragile resource, supersensitive to environmental conditions. Its future success depends first of all on maintaining the high quality of Lake Superior's cold, clean waters.

The lake trout is also extremely vulnerable to recreational angling, which, unlike commercial fishing, cannot be as rigidly controlled. Today's booming recreational fishery for lake trout may also have to be fine-tuned, with increased and more complex regulations.

Despite an apparent success, the lake trout's position as top predator remains uncertain. In pre-lamprey days, the trout depended upon an intricate yet stable system of forage production. The lake herring and chubs were the main elements of that system. With major changes in these forage fishes—including some losses of species—the lake trout's food changed to nonnative, or exotic, species, such as smelt. But population levels of exotics are seldom stable, and future production of suitable forage fishes also remains uncertain.

The future of the lake trout along the North Shore will continue to depend upon intensive research, massive hatchery culture and plantings, and indefinite control of the sea lamprey. Above all, success will require a sensitive understanding by scientists and the public alike of the problem we face. The final solution has not been reached; there is no quick fix; greater knowledge and sacrifice will be necessary. Those of us who appreciate *Salvelinus namaycush*—whether on the end of a line or on the table, or simply as a great natural resource indigenous to the inland sea—believe the effort is worthwhile.

The native species of *Coregonus*—lake whitefish, herring, and the five chubs—also underwent drastic changes. Heavy fishing prior to the arrival of the sea lamprey, as well as lamprey predation itself, contributed to these changes. Commercial fishing efforts were redirected toward lesser species when populations of lake trout and whitefish declined, causing decreases in the numbers of these smaller species. Some forage species, for example, the bloater, became more abundant when predation by lake trout was relieved.

This situation was a classic example of the intricate change that occurs when a finely balanced community of predators, competitors, and prey is upset at one point.

Largely relieved of predation by sea lamprey and supplemented by planting of hatchery-reared fish, whitefish today are making a comeback parallel to that of lake trout. Within a decade of lamprey control, whitefish catches approached levels near those prior to the lamprey invasion.

The herring is the most abundant native fish and largely supports the commercial fishery of today in Lake Superior, especially in Ontario's North Shore waters. And, of the five chubs originally in Lake Superior, only the bloater remains for all practical purposes.

Exotic species of fish began arriving in Lake Superior shortly after white settlement upon its shores. Introduction had been made by accident—usually through human activities—or by deliberate actions. By the beginning of the twentieth century, many deliberate introductions had been made almost throughout the continent; stocking of new species was, in fact, considered at that time one of the most beneficial fisheries management tools in the state-of-the-art. The role of exotic species— whether as new predators, prey, or competitors—remains uncertain and unstable.

Few records remain of early introductions. Brown trout (*Salmo trutta*) were brought from Europe in the late 1800s and released in many stream systems in the eastern United States; rainbow trout were similarly relocated from the Pacific coast of North America about the same time. The rainbow (*Salmo gairdneri*), introduced to Lake Superior in 1895, was especially successful and established self-sustaining populations on the North Shore within ten years. Later the rainbow itself suffered from sea lamprey depredation, but like the lake trout it is now on the increase. Both the rainbow and brown are highly regarded and fill important niches in recreational fishing today, although the brown remains at low populations in Lake Superior.

163

At one time or another, almost every salmon, Atlantic and Pacific, was tried in the Great Lakes, but all failed. We think today that the reason for these early failures was that the native lake trout-whitefish-herring-chub system was so well established and stable that nonnative species could not penetrate it.

But that native stability was lost when heavy fishing exploitation, followed by sea lamprey predation, disrupted the system. New predators now successfully share the position of top predator with the lake trout and burbot; new forage species replace the old as food for the predators. The system is more complex now, and more difficult to understand, for the newcomers have different environmental requirements with respect to such factors as water depth, temperature, food, and spawning needs. As with the lake trout and whitefish some of the newcomers' populations are maintained or supplemented by hatchery plants, and these stockings vary almost from year to year, as new techniques and ideas are implemented.

At this point, we may hypothesize a scenario something like this: The sea lamprey decimated the lake trout and the burbot, the only major fish predators. Since the trout had historically preyed upon the herring as its principal food, it would be expected that the herring population, relieved of the trout's predation pressure, would respond with a dramatic increase. As it turned out, the lake herring did not increase.

Another species—the rainbow smelt, *Osmerus mordax*, native to the Atlantic Ocean—had arrived several decades earlier but had been slow to expand. Smelt eggs were stocked experimentally in the St. Marys River in 1909 and some later years, and the species had been planted experimentally in an inland Michigan lake in 1912. From one of these sources—we aren't sure which—the smelt made its way into Lake Superior and was first observed there in 1930.

Before the lamprey came, lake trout had fed upon the smelt as well as herring, apparently holding down the smelt's population. But with the relief of the trout's predation, smelt populations did increase—exploded, actually—and the halcyon days of smelt dipping on the North Shore in the 1950s and 1960s became legendary.

Now it appears that the smelt, an exotic with the capability of rapidly exploiting a new habitat, is also a predator of sorts. Although it is a small fish, it has teeth, and it apparently fed heavily on herring fry, well before herring could spawn, holding down—even decreasing—herring numbers. Add to that some lamprey predation on herring (but not on the smaller smelt), and we can perhaps understand at least in part the segments in the lake trout-herring-smelt succession—and why herring failed to meet the expected increase.

Today, smelt are on the decline, although the reasons for this are not entirely clear. Increased lake trout populations may again be functioning to control smelt. Or, as new hypotheses may explain, the pink salmon, another exotic, might be preying heavily on smelt fry. The numbers of pink salmon in Lake Superior are now declining, but we don't know exactly why, nor can we predict what effect this development will have on smelt.

In 1956, about 21,000 fingerling pink, or humpback, salmon, destined for stocking in Hudson Bay tributaries far to the north, were inadvertently released instead into the Current River, at Thunder Bay. Of all the exotic salmons, Atlantic or Pacific, the pink salmon was regarded as the least likely to survive and become established in fresh water. But in 1959, two spawning pink salmon adults, progeny of the fish spilled at Thunder Bay, were taken by anglers from North Shore streams, one each from Minnesota's Cross and Sucker rivers.

The life history of anadromous pink salmon is almost strictly on a two-year cycle, and in odd-numbered years following 1959 other spawning adults were recorded in Lake Superior tributary streams. During the rapid changes taking place in the 1960s, the pink's population exploded, so that in the odd-numbered years of the 1970s, the abundance of this species reached such high levels that it was considered a nuisance. Pink salmon peaked in 1979, but, for unknown reasons, may now be declining.

Varying slightly from its two-year cycle, the pink has now established stocks that spawn in even-numbered years, first recorded in 1976 in Ontario's Steel River. The pink salmon's naturally reproducing capacity in North Shore streams has been great; no stocking other than the original accident has been made. By 1979, pink salmon had dispersed from the North Shore of Lake

Superior into all the Great Lakes. It must be considered today as a firmly established resident of the Great Lakes.

The pink is small in Lake Superior, reaching a spawning size of only about one pound to one and a half pounds, about fifteen inches long. Although it is difficult to catch by hook and line, expert anglers have developed techniques to take the pink in spawning streams. The mature male develops a fleshy, humped back, leading to this species' other common name. The pink salmon starts to deteriorate rapidly when it enters a stream for spawning, and soon develops open, fungus-covered lesions and a quickly blackening body. However, when bright and fresh from cold North Shore waters, silver in color, the pink is delectable table fare. Of its ultimate place in Lake Superior's fish community, however, we are uncertain.

Two other exotics—both large and piscivorous, both Pacific salmon—have made major impacts upon Lake Superior's fish system: the coho salmon and chinook salmon. They now are well established as principal elements in the lake's recreational fishery and in the other Great Lakes as well. Their establishment was met by an initial dramatic success—and a longer period of fine-tuning, introspection, and some changed priorities.

When it became so abundantly clear in the late 1950s that Great Lakes fish populations had suffered a tremendous upheaval—especially with enormous alewife numbers developing in Lakes Michigan and Huron—the search was on for some other species to supplant lake trout as predators in order to utilize the abundant forage fish. Although Pacific salmon had been tried before, and failed, long before the arrival of the sea lamprey, fishery managers and administrators felt the drastically changed circumstances now might permit successful introductions. With lake trout and burbot essentially gone, no other predators existed, and the alewife, in places at great nuisance levels, provided a superabundance of possible forage. Likely candidates were the coho and chinook, the largest of the Pacific salmons. Their introduction into Lakes Michigan and Huron were immediate successes, and the state of Michigan, located between these two Great Lakes, caught salmon fever.

The success of salmon plantings was so heady that similar introductions were made in Lake Superior, too—coho in 1966, chinook in 1967—by the State of Michigan. The first coho salmon

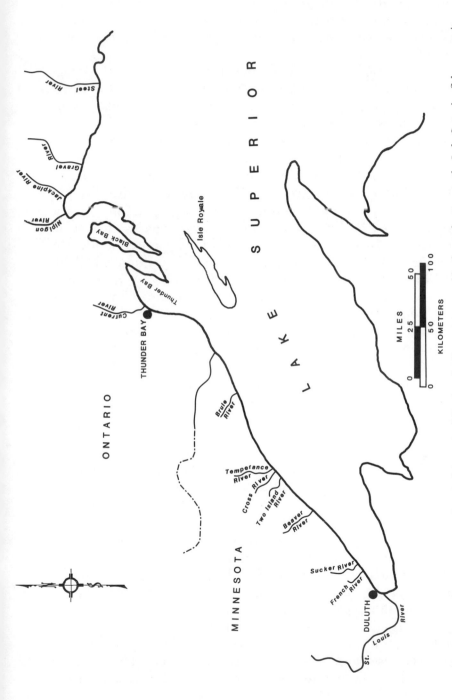

Sites important in the introduction of exotic fish species following the disruption caused by the sea lamprey to the Lake Superior fish community. Most introduced species were anadromous—river-running to spawn—and thus were released and subsequently caught in tributary streams.

plantings on the North Shore were in the period 1968-70, by Minnesota and Ontario, in Two Island, French, Beaver, and Brule rivers (Minnesota) and in the Jackpine and Gravel rivers (Ontario).

But Superior's waters are much less fertile than the other lakes, its water temperatures lower; and the alewife had not proliferated, to provide forage. So results were not nearly as spectacular as in Lakes Michigan and Huron. Still these latest newcomers were successful in adding major components to Lake Superior's fish community and to an increasingly diverse sport fishery.

The coho and chinook, like all salmon, are anadromous, ascending tributary streams for spawning. However, natural reproduction in the short and relatively few spawning streams on the Minnesota coast has been unsuccessful, or in any event insufficient to add substantially to lake populations. Consequently, the maintenance of large lake populations has depended on plantings of hatchery-reared juveniles of both species. Many millions of coho and chinook have been stocked in American waters since the late 1960s, mostly on the south shore.

Perhaps more mindful of the emphasis on rehabilitating the lake trout as the principal large native species, Ontario planted small numbers of coho only in the early period 1969-71, and since then no chinook or coho at all. However, both coho and chinook appear to be spawning successfully in Ontario streams, particularly on the eastern shore.

Of the two Pacific salmons, Minnesota continues to stock only chinook now, having placed about two million young chinook salmon in North Shore waters in the past two decades, maintaining a highly successful recreational fishery.

For whatever reasons, some of them still held largely secret in the depths of Lake Superior's waters, the long-range trend in North Shore fish populations is in the direction of restoration of the native fish complex. The lake trout, whitefish, and herring all appear to be on the way back.

Interestingly, herring populations in the Lakehead area—Thunder Bay, Black Bay—never did suffer the drastic declines experienced elsewhere along the North Shore. Possibly the smelt in the Lakehead waters, for subtle environmental reasons, never attained a strong predator position over herring. In any event, the herring in this part of the lake is the largest commercial fishery along the North Shore today.

The five native species of chubs: what happened to them? Only one—the bloater, smallest of the five—is left. Here we know even less of what happened. We do know that declines in the other four species were experienced prior to the time the lamprey arrived in Lake Superior, the apparent result of heavy fishing. With the loss of long-established predators, and introduced new predators and prey, intricate relationships among the chubs must have changed in subtle ways.

We know that the five chubs, very closely related, hybridized easily. The bloater of today is much larger than the bloater of pre-lamprey days—certainly a hybrid of some sort. For the moment, at least, even this event has its advantage. For the bloater, in sizes now making commercial fishing profitable, taken from the cold waters of Lake Superior, smoked, and sold in North Shore stores and roadside stands—"smoked chub"—is delicious eating.

The alewife, another Atlantic Ocean native, apparently entered the upper Great Lakes through the Welland Canal, along with the lamprey, arriving in Lake Superior in 1953. Enormously successful in Lakes Huron and Michigan, the alewife became a nuisance of great proportions; Lake Superior, however, is cold and outside the alewife's temperature tolerances. And although a few alewives are occasionally observed or taken in gill nets, the North Shore was spared the worst of the alewife nuisance.

The lamprey and alewife made their entry into the upper Great Lakes through our modern canals and were able to colonize their new freshwater environments rapidly. But why not also other marine species? The Atlantic Ocean is rich with many species in its fish fauna. Are there other fishes, and perhaps also invertebrates, that for some unknown reason are only able to navigate through locks and waterways slowly, and which may yet invade the Great Lakes? An intriguing question, indeed. It is a situation that bears watching.


Now that certain exotics have become so successfully established on a self-sustaining basis—the sea lamprey, smelt, pink salmon, possibly the coho and chinook—and certain others have essentially been exterminated—sturgeon, some chubs—it is apparent that to completely return Superior's fish community to its pristine condition is clearly impossible.

Many problems stand in the way, perhaps the greatest being the sea lamprey. Complete elimination of the lamprey is remote, and concern about continued use of chemical control remains. However, other water bodies hold both the lamprey and the lake trout in apparent stable coexistence; perhaps once other native fishes in Lake Superior are brought again into an ecological balance, with at least the major species returned to something like their original condition, the same happy situation might yet be attained on the North Shore. Some leading fishery scientists and students of Lake Superior have expressed, cautiously, the opinion that this goal can, with perseverence and sacrifice, be reached.

# ELEVEN

# The Fish Market

The smooth, pendular swells of Lake Superior, originating far to the southwest, rolled the little watercraft in an undulating, circulating motion. Deftly, the stocky fisherman pulled the gill net over the gunwale, his movements matching the rhythm of the swells.

An assistant, at his oars, kept the boat in alignment so that its movement neither overran the net nor itself pulled on the net. Beneath the boat, the green, rocky bottom was visible through crystal-clear water, for the lake was shallow here, about twenty to thirty feet deep. The boat and its two occupants were only a hundred yards from the sloping, rocky shore.

Then, from the emerald depths fifty feet away, appeared the slim, white form of a large fish, its body flashing with light as it struggled in the net. The fisherman heaved the fish up and over the gunwale. With quick fingers, he loosened the fish's gill covers from the entangling, heavy linen strands. The fish fell into a box on the boat's deck—a lake trout of some ten pounds or more.

With a few more rhythmic pulls on the net another white form emerged in the water out from the boat, and then, in rapid succession, two more trout close together. Soon the box began to fill, and the assistant readied another.

The place was along a rocky shore near the mouth of the

171

*Commercial fishermen, Grand Marais*

Pigeon River, and the fishermen were employees of the American Fur Company, operating out of the Grand Portage post. The season was late September, and trout were abundant in nearshore waters.

The year: 1836.

Much of the fishing enterprise changed in the ensuing 150 years. The boat used later was a great deal larger and powered by steam, rather than by sail or oars and muscle. Gill net construction evolved from linen to cotton twine to monofilament nylon strands, much more efficient at entrapping fish. Instead of handling the heavy gill nets with muscle power, the fishermen operated mechanical lifters that paid out and retrieved longer gangs of nets over power-operated drums. No longer was a single assistant sufficient for the larger craft or for the longer trips farther from shore, and operation of mechanical equipment required many more hands.

But the greatest changes during this period evident today—were toward less abundance, smaller fish, and fewer fishermen.

Overall, the commercial fishery collapsed, owing to direct and indirect effects of intense fishing and sea lamprey predation. The number of fishing operations was reduced drastically. Seasons for taking lake trout were shortened, numbers of fish taken tightly regulated, and then marketed largely only locally. Most commercial fishing effort is now directed toward other species: whitefish, herring, chubs, smelt.

With the current boom in recreational lake trout angling and its much greater economic value, it now appears unlikely that commercial fishing for trout will again be a mainstay of industry in Lake Superior.

Yet the history of this century and a half of commercial fishing along the North Shore makes a fascinating tale, rich in daring and struggle by settlers of the coast. For it was the fish resource of these cold northern waters that fed the itinerant fur traders, lumbermen, and copper and iron prospectors, and brought about permanent settlements.

The prehistoric Ojibway living around the east side of Lake Superior fashioned gill nets with strands of the inner bark of willows. They fished the nearshore waters from their birchbark canoes in summer and fall, through ice in winter. Additional fishing with baited hooks and spearing at night by torchlight also brought them trout, but use of the gill net, at which the Indians were highly skilled, provided them with the bulk of their catch.

Whereas fishing provided food for families and bands, probably through cooperative efforts in summer fishing villages, it seems unlikely that the Indians did not at some times engage in trading activities involving fish, perhaps between bands or other villages. Of this earliest commercial fishing, nothing remains on record.

After Europeans made contact with the Ojibway, ample written evidence was recorded on the use of fish in trade. The Indians bartered freely with the explorers and fur traders, exchanging fish as well as beaver pelts for European goods. Little remains in the record of quantities of fish caught or exchanged, so that we have no reliable data on fish abundance or catch. But the quantities greatly impressed the Europeans; their accounts were expansive.

In the early 1800s, when the fur trade era was near its peak, fishing stations operated by whites were developed at several points around Lake Superior. The major ones on the North Shore were at Sault Ste. Marie, Grand Portage, and Isle Royale.

The American Fur Company, finding its fur trading enterprise on the decline, in 1835 began the first commercial fishing industry of major importance on the North Shore. The company established a major post at Grand Portage in 1836, five stations on Isle Royale the next year. The main headquarters were located at La Pointe, Wisconsin, on Madeline Island. Soon the company's profit from fishing exceeded that from its traffic in fur.

The company had launched the sailing schooner *John Jacob Astor* in 1835, and it presently began carrying fish. Two more schooners were soon added to the company fleet. At the North Shore fishing stations skilled coopers made barrels, carpenters made fishing boats; whitefish, lake trout, and siscowet were salted in the barrels. The company's boats picked up fish for inspection and repacking at La Pointe, and then fish were shipped to the Sault, and on to lower ports. Shipments were made to Chicago, Detroit, New York.

With modern entrepreneurial ardor, the company attempted to develop even greater markets in the Mississippi Valley through Illinois, Indiana, Ohio, and the cities of St. Louis on the Mississippi and Louisville on the Ohio, even as far south as New Orleans. Production reached levels of 2,000 to 3,000 barrels per year.

As the American Fur Company thus expanded its fishing and markets in the late 1830s, the Hudson's Bay Company took alarm. The American company even operated a fishing station at the Montreal River, well into Canadian territory on the eastern side of the lake. At first the older company keenly felt the competition from the Americans for Indian labor; Indians were thus diverted from fur trapping, especially in the Fort William area, not far from Grand Portage. And so the Hudson's Bay Company entered the commercial fishing business, too, largely to counter the Americans' threat. Major Hudson's Bay fishing stations were developed at established fur posts—Sault Ste. Marie, Michipicoten, Pic River, Fort William—entering the commercial fishing enterprise in 1839. Many seasonal outposts were maintained by the major stations; seventeen were supported by Fort William alone.

Whitefish and lake trout again were the major species, but other fishes, such as suckers, sturgeon, northern pike, and walleye, were also taken. Fish were mainly preserved by salting—"put into pickle"—and some were iced down with Lake Superior ice. Some Hudson's Bay fish were even shipped to United States markets down the lakes, and farther east to Montreal.

The American Fur Company's attempts to establish a national market for Lake Superior fish throughout the states soon failed. The Panic of 1837 depressed trade; distant markets were not successful. Except for a few boats operating around Isle Royale, fishing was dropped in 1841, and in 1842 the American Fur Company folded completely. But the company's enterprise on the North Shore had marked the first attempt at a major commercial fishery in the Great Lakes.

In midcentury, small, temporary settlements developed along the American portion of the North Shore, and many accounts of fishing give evidence of highly productive operations, providing fish locally. The abundance of lake trout, whitefish, and herring—with no apparent end in sight over the vastness of Lake Superior's waters—gave rise to expressions such as "inexhaustible," "exhaustless," "rich and unlimited," and "The demand can never exceed the supply."

In the late 1850s, commercial fishing boomed in American waters. The Ojibway land cession treaty of 1854 spurred a rush in mineral prospecting, with the consequent increase of local markets for fish; the treaty also permitted the establishment of permanent fishing stations and settlements.

Distant markets, however, were largely denied to these early fishermen, because of the lack of suitable transportation. But the first government road beween Lake Superior and St. Paul, Minnesota, opened in 1856, and then frozen fish could be transported over this road in winter to southern cities by dog team. The years 1856 and 1857 brought great numbers of prospectors and land speculators, more established villages—and greater local markets for fish.

In addition to the whitefish, lake trout, and herring fishing, primarily in autumn, an important effort was applied in springtime to walleye and northern pike in the estuary of the St. Louis River. Early records of this fishery are confused as to species caught, however, for such terms as *pickerel*, *pike*, and *wall-eyed pike* were interchanged locally. The inaccurate term *wall-eyed pike* persists today.

About this time, too, brook trout were taken by angling in the many tributary streams along the coast. In higher numbers and larger sizes than today, the brook trout was taken mainly for sport, as it is today. The delicious fish was highly prized, although it probably never entered into the commercial fishery.

During the American Civil War years, mineral prospecting increased along the North Shore, and so did settlement. Fish were brought to the Superior, Wisconsin, port at the western end of the lake, shipped down the Great Lakes by sailing schooners through the Sault to lower lake ports and eastward, and also across Lake Superior to upper Michigan, as well as sold locally. Still there remained the greatest obstacle of a rapid, all-season link to southern markets and to expanding frontier settlements across the nation.

Then, in 1870, the completion of the Lake Superior and Mississippi Railroad, linking Duluth and St. Paul, removed this obstacle and brought about a greatly expanded market for Lake Superior fish. Commercial fishing increased again with the improved distribution system, and citizens in the southern Minnesota cities and along the Mississippi were treated to fresh trout and whitefish from the cold waters of Superior. Established fishing operations along the Minnesota coast increased in number and size, including new stations at Beaver Bay, Grand Marais, Brule River, Pigeon River, and points in between.

The commercial enterprise was facilitated by the relative ease with which a hardy immigrant could get started in the fishing industry. With only a small capital investment in a boat and gill netting, one man could start up a profitable fishing business. Often aided by a son or brother as partner in his small boat, a hardy Norwegian, new to the American land of opportunity, was in business. Since most fishing could be productively conducted near shore, small boats and little equipment were sufficient.

The closing decade of the nineteenth century was the period of greatest proliferation of commercial fishing operations on the Minnesota coast. Norwegian immigrants, toughened to the rigors of fishing off the cold, rugged shores of their native land, felt at home on Lake Superior's northern coast. They sent back glowing accounts of new opportunity, and soon more families arrived. In 1890, only about fifty fishermen were active, but by 1917, after the main immigration, there were 273 license applications along the Minnesota shore. Families and fishing operations, with their log cabins, fish houses, and net-drying reels, soon filled every little bay and cove from Duluth to Grand Portage. Soon something like 400 individual fishermen occupied the shore, continuing into the 1930s.

Fishing boats, gear, and techniques developed rapidly to serve the needs of the rapidly growing numbers of family-type fishing operations in Minnesota.

The basic fishing craft was the skiff. About fifteen to thirty feet long, flat-bottomed and uncovered, it was a simple, utilitarian craft that served the early fisherman well. Homemade of native wood by the fisherman, or by a local carpenter in the community, the skiff was cheaply and quickly produced. It was built double-ended—sharp bow and stern—for seaworthiness; this construction allowed either end to rise on high waves without shipping water, or when the boat was fast to a net and was being pounded by heavy seas. A partial bilge below floorboards kept water from feet, clothing, and gear. Provision for simple sails sometimes meant relief from miles of rowing; later, one end was squared off to hold an outboard motor. The skiff could hold up to 1,000 pounds of fish.

The larger Mackinaw boat, still open but sometimes with one or two sails, meant even greater savings in effort. Like the skiff, the Mackinaw boat was flat-bottomed, with even higher pointed bow and stern. Adapted, apparently, from the style of the *canot du nord* of the Indians, the Mackinaw boat developed into the most used fishing craft of its time on Lake Superior, famed for its seaworthiness. It permitted longer, safer trips offshore, greater loads of fish.

With steam engines, boats were built larger and then covered. The increase in boat size, however, was limited on the Minnesota shore by the lack of safe harbors; all boats, for protection from storms, were hauled out of the water upon landing. So size of boat, whether skiff or tug, was restricted to a length that could reasonably be pulled up—about thirty-five feet. All fishing locations had a log ramp and capstan for hauling up boats.

Such a restriction on size was not necesssary on the south side of Lake Superior, so larger boats were developed there. To the consternation of North Shore fishermen, these larger outfits sometimes operated in northern waters. In the 1930s, Minnesota fishermen successfully persuaded the state legislature to pass a law restricting boat size in Minnesota waters—to thirty-five feet!

Before the crash in fish populations came in the 1950s, the modern diesel-powered cabin boat appeared. It was a highly efficient industrial operation—setting miles of gill net, far from shore, pulling nets with power-operated lifters, requiring a crew of hands—and capable of holding tons of fish. Often, on the long trip back from far-off shoals and islands, fish would be cleaned, iced, and packed in boxes before reaching home port.

In the last two decades of the 1800s, fish transport and marketing systems developed into efficient and profitable routines. A. Booth Packing Company of Duluth operated the steamer *Hiram R. Dixon*. Making the round trip three times a week, Duluth to Grand Marais or Grand Portage, the *Dixon* picked up fish and unloaded provisions and passengers, even livestock. In a small community lacking a deepwater dock, fishing citizens were forced to relay fish, goods, and people by skiff between shore and steamer—a hazardous job in storm or at night. Horses and cattle were made to walk the plank, and they swam ashore.

It was a point of civic pride for a fishing community to develop deepwater facilities for the steamers with rock- or log-cribbed wharves and breakwaters.

The *Dixon* was retired in 1902, ran aground on Ontario's Michipicoten Island later the same year, and was destroyed by fire the next year. After that, Booth operated the *America* similarly, including runs to Isle Royale, until that steamer was wrecked in Isle Royale's North Gap Channel in 1928.

Other steamers and powered boats also served the Minnesota North Shore in these years: the earlier *Charley* ran up the shore to Beaver Bay, the *Bon Ami* to Tofte. The "Mosquito Fleet" of eighteen gasoline launches was operated by fishermen out of Two Harbors in the years 1907-24. Operating from Duluth, other fish companies included Sam Johnson & Sons, H. Christianson & Sons, Hogstad Fisheries, Scandia Fish Company, Bray Fish Company, Goldish Fish Company, and Sivertson Brothers.

Only the last of these, Sivertson's, survived the collapse in the 1950s and is still operating. Beginning in 1892, Sam Sivertson and his wife, Theodora, fished for Booth and worked the Isle Royale waters. Two sons, Stanley and Arthur, grew up in the fish business and continued as Sivertson Brothers, starting in 1938. Today, Stanley Sivertson's fishery continues—as Lake Superior Fish Company (purchased from Goldish) in Duluth, and Sivertson Fisheries in Superior, Wisconsin.

With the new railroad from Duluth to St. Paul in 1870, fish were railed to St. Paul and on to more southern markets. The system of fish pickup by powered boats and steamers was phased out in the 1920s, however, when Minnesota's North Shore road was completed, and thereafter the bounty of Lake Superior's waters was hauled to Duluth by truck.

The working part of the fisherman's business was the gill

net. Up to 500 feet long, one net often would be fastened to another, or more, to form a *gang*. A gang of ten nets, fished from a tug, might be a mile long, and gangs even longer were often set, but a gang of two nets was about all one man could handle at a time in a skiff. Later, large powered boats and net lifters allowed longer sets. Today, restrictive regulations do not permit the long gangs of yesteryear.

To the top line of the gill net were fastened wooden floats, and along the bottom line lead weights were fastened every six feet or so. Between the two lines, no matter what the depth at which the net was set, the gill net itself hung vertically, to intercept moving fish. Floats were hand-carved from local woods, usually white cedar, and the leads were made by a local smith who had the facilities for melting, pouring, and molding the lead weights. At the end of each gang, an anchor held the net in place. A buoy with flag—sometimes made of the fisherman's worn-out shirt or underwear—marked the net's location.

For the bottom-ranging lake trout, gill nets were set on the bottom in relatively shallow, nearshore waters. Trout nets were constructed about six feet deep—that is, in a vertical dimension when set. These were mobile nets, moved frequently to follow the fish; anchors and buoys were lifted and moved each time. Lake trout were fished deep in summer but toward shore in fall, although this varied with the stock and location. When shallow, nets were often set with one end almost up on a rocky shore. With a lack of large, flat shoals along the Minnesota coast, whitefish were taken rarely. Lake trout, although not as abundant as in Canadian waters, predominated in these nearshore catches. Only this narrow strip of fishing shoals suitable for lake trout was present, since water depth increased quickly offshore—to as much as 700 feet within three miles from shore. Consequently, Minnesota waters produced neither lake trout nor whitefish as prodigiously as the more abundant shoals on the eastern side of the lake.

On the other hand, Minnesota waters predominated in their herring catch. Fishing for herring was conducted farther from shore, where these pelagic fish congregated in vast numbers over deep waters. Lake herring are sometimes called shallow-water ciscoes, because of this habit of occupying upper strata of water, even if over great depths.

180

Herring nets were constructed with smaller mesh and a greater vertical dimension, over ten feet. They were set neither on the bottom, as were trout nets, nor floating at the surface, but rather at some midwater depth, depending on fish location. The depth of set was changed up or down with change in fish depth, determined by the fisherman's recent experience. The anchor at each end—usually a 100- to 200-pound stone—was permanent. Too heavy to retrieve, an anchor stone was cut loose from the anchor line at the close of a netting season, to be replaced with another next time.

Good anchors for herring sets were at a premium. They had to be large stones with a shape that permitted the tying on of line; the ideal was a sort of dumbbell shape. Lake Superior's rocks being what they are, worn round and smooth, good gill net anchors were hard to find. In the absence of proper anchor stones, the fisherman might be forced to chip away at the sides of an oval-shaped rock to create the notches necessary to receive a tied line.

The tale is told of a fisherman who, happening upon a cove with many large rocks of just the right shape, considered he had found himself a rare treasure trove. Being unwilling to leave his find unexploited, he filled his skiff with his treasures. With lake waters lapping over the gunwales, he rowed for home. But within sight of his home dock, the skiff sank, and anchor rocks and skiff as well went to rest on Lake Superior's bottom.

No doubt, thousands of dumbbell-shaped stones remain today on rocky bottoms off the North Shore in Lake Superior's cold depths, part of our archaeological heritage from the heyday of commercial fishing.

In lifting a herring net, the flags were first located—sometimes by instinct when in a dense fog—and the end of the net made fast to the skiff. Then the net was lifted over the boat and, with floats, leads, fish and all, pulled across the gunwales. The fisherman literally pulled the boat along under the net. He would remove a herring by pushing the fish through the entangling mesh; thus, herring nets were known as "herring chokers." By pulling the net over one side of the boat, then "spinning" the net back into the water over the other side, the net was reset. Care had to be taken to ensure that the net was returned to the water without being twisted or tangled. When the fisherman reached

181

the other end of the gang, his catch was in the boat, and the nets were ready for the next day's fishing.

Whereas most herring fishing took place far from shore, in the fall herring ran into nearshore waters for spawning. Fishing was good to time of freeze-up, and later. Herring catches were so good, so late, that gill netting was continued under the ice, a cold and sometimes dangerous activity.

Obviously, techniques were different in winter. Through clear ice a hole was chopped, and a long pole with line attached— usually a trimmed, slim tree trunk—was pushed through the hole. The progress of the pole could be followed under the clear ice, and at its end another hole was chopped. The pole was pushed again to its end, trailing its line. When sufficient distance was attained, after many holes and pushes, the pole was removed, and the line was used to pull a gill net into position. Markers set into the ice at either end located the net's position for retrieving the catch the next day.

Picking thousands of herring from the fine strands of the nets, amid Lake Superior's wintry blasts, was a finger-numbing job. But catches were high, and herring were frozen on the spot, ready for boxing and shipment.

Late winter was the time for skiff building, making fish boxes, and mending nets.

Weather, of all kinds, was a constant factor in the life of a North Shore fisherman. The storms of autumn—when fishing was best—are legendary. On the Minnesota coast, trending northeast-southwest, the nor'easters of November blew parallel to shore in harrowing gales, ripping up docks, fish houses, boats, and net-drying reels.

Out on the lake, however, it was the sudden storm from the *northwest* that was feared most. Arriving unexpectedly over the shore's highlands, blasting straight out from the protecting shore, it could catch a fisherman in a small boat with insufficient warning and blow him out onto Superior's open waters. If he was able, and lucky, a fisherman tied his boat to the nets and rode out the storm, to return to shore when the blow was over, possibly the next day, after a terrifying night.

The chilling temperature of Lake Superior's waters, in any season, meant that a capsize and dunking, without help, were not survived except for a brief time.

Storms were so severe sometimes that nets would be torn from the anchors in the choppy waters, a serious economic loss. Or bad weather would prevent a fisherman from tending his nets, for possibly a week or more, with consequent loss of fish to spoilage.

The likelihood of fog, too, was an ever-present danger. And fog-squalls could move in rapidly. Without compass, or sun or stars, a man and boat could be hopelessly lost on the big lake. Then instinct was relied upon, an attribute of the experienced Norwegian fisherman that was sometimes pridefully trusted. But sometimes, too, instinct was faulty, and more than one fisherman found himself far out on the lake when the fog lifted.

Despite grueling work, frozen fingers, and the hazards to life that Lake Superior constantly presented, the tough immigrant fishermen from northern Europe loved the shore and the fishing life. Once they had established their homes and families, and tasted the freedom of newfound opportunity, few became discouraged and abandoned the North Shore.

Many are the fishermen who have recounted their simple and honest appreciation of the rocky shore's challenge. Perhaps, among all the North Shore's pioneers, it was the commercial fisherman who most respected and esteemed the changing temperaments of the great lake, the beauty, freedom, and solitude—and danger—on the open water or along the wild, spruce-lined shores.

To these Scandinavian individualists, the North Shore was like home.


Commercial fishing along the Canadian portion of the coast developed later than in American waters. As in Minnesota, early commercial enterprise had been established mainly to serve the fur trade, especially in the Michipicoten area, at the Pic River post, and, after the move by the North West Company from Grand Portage to the Kaministiquia, at the company's Fort William center. Fishing served the fur trade until after the mid-1800s, when the fur era came to a close.

Analogous to Minnesota's walleye fishery in the St. Louis River at the western end of Lake Superior was the Indians' whitefish fishery in the rapids of the St. Marys River at the eastern end.

The fishing attracted tribes from Wisconsin, James Bay, and Lake Nipissing. To these early Americans, the whitefish was regarded as the "caribou of the water"—their second most staple food. Many accounts of this fishing activity are recorded, among the earliest being that by the explorer Radisson, in 1659. Apparently only the Indians were able to master this technique of whitefish fishing, for it required such long-developed skills in handling the birch canoe in turbulent rapids that the white man could never acquire the necessary proficiency.

With two men in a canoe, an Indian fishing team started at the upstream end of the rapids and drifted down. The bow man, armed with a scoop net attached to a long pole, would lay the hoop of the net on the bottom of the stream, *over* the fish, usually in an eddy or pool formed below boulders. Then, with a twist of the pole, as the stern man with paddle allowed the canoe to drift downstream, the netter brought the net into the canoe, heavy with its load of whitefish. Fish were large, six to fifteen pounds each. The paddler would return to the eddy, if he could, for another effort in the same pool, or drift down to a new pool. The operation required not only skilled canoemanship, but adept coordination between paddler and netter. Upsets were common.

Termed *Saulteurs* ("people of the rapids") by the French, the native Ojibway had fished whitefish in great quantities from the St. Marys in prehistoric times for their own use. And when Europeans discovered the Indians' activity, their impressions led to extravagant estimates of this great resource, including the statement that the fishery alone could support 10,000 persons.

In the 1700s the rapids fishery served the fur trade around the Sault, and in the 1800s whites exploited the Indians' harvest in commercial trading.

Later, as the rapids were impounded and canalized by the development of navigational locks in the later 1800s and early 1900s, the ancient whitefish spawning habitat was obliterated, and the famed fishery came to an end.

In the latter half of the nineteenth century, major changes took place along the Canadian coast, greatly affecting the trade in Lake Superior fish. With the close of fur trading activity, local needs and markets for fish declined, and so did Hudson's Bay Company fishing.

But land cession in the Superior-Robinson Treaty of 1850 spurred settlement and, to an even greater extent, mineral prospecting and mining. Logging and lumbering also generated needs for locally produced food.

In the 1880s, an upsurge in commercial fishing occurred along the Canadian shore. Major changes in boats and equipment resulted in improved efficiency in harvesting fish and in a much greater overall catch. Rowboats with one man were replaced by small sailboats with a crew of two or three. Steam-powered fishing vessels, called "steam tugs" by the fishermen, appeared on Lake Superior in the 1870s, permitting transportation to distant markets. Tugs became common in the first decade of the new century; and the old days of oar and sail were over.

In the 1890s, steam-powered net lifters appeared, and the drudgery of hand-hauling gill nets ended; the powered lifter also permitted longer gangs of gill nets, further increasing the intensity of fishing effort. Later, gasoline engines replaced steam, and later yet diesel engines were to largely replace gasoline. Operation of the steam tug, up to ninety feet in length, was a highly organized effort. No longer was a crew of fisherman and assistant sufficient; a crew of five men or more captained and engineered the boat and also handled nets and fish.

By far the major innovation in gear—changing the materials used in gill nets—came along during this period of intensive fishing. Linen strands were replaced by cotton about 1930, but this improvement was slight compared with the later shift to nylon in the 1940s. Lighter, thinner, less visible to fish, nylon gill nets were estimated to be up to three times more efficient in taking lake trout.

Of most importance, however, as in Minnesota, was the development of long-range transportation later in the 1800s, which allowed the shipment of fish to distant markets. With the completion of the transcontinental Canadian Pacific Railway in 1885, fresh, iced whitefish and lake trout could be marketed in eastern Canadian cities, such as Toronto and Montreal, as well as in American cities on the lower lakes.

In addition to shipping fresh fish, the railroads bought fish to serve in their dining cars, a luxurious treat for transcontinental travelers along the North Shore.

And so fishing boomed in the first half of the 1900s. The halcyon years along the Canadian coast were 1913 to 1919, with the peak in 1915. Fishing stations were established from Fort William and Port Arthur in the Lakehead area along the coast all the way to Sault Ste. Marie. Major stations were Port Arthur, Nipigon, Rossport (formerly McKay's Harbour), Port Coldwell, and Quebec Harbour on Michipicoten Island. Other depots were located at Hurkett, Jackfish, Heron Bay, Pic River, Otter Head (now in Pukaskwa National Park), the Mission at Michipicoten, Gargantua Harbour, Sinclair Cove (now in Lake Superior Provincial Park), Agawa Bay and Montreal River, Mamainse Harbour, Batchawana Bay, and Goulais Bay, as well as at other remote points.

The Nicoll brothers' operation at Port Coldwell was especially noted for supplying fresh lake trout for the dining cars of the Canadian Pacific Railway, a custom still maintained on present-day VIA Rail, although fish do not now come from off Port Coldwell. The Nicolls' company did business for forty years, starting in 1915, employed as many as thirteen men in its own operation, and bought fish from local independent fishermen as well. The company fished several steam tugs and smaller craft. Each tug carried enough nets for a set up to seven miles long, with each single net, or "piece," extending 600 feet. Steam-powered net lifters were available in these years, without which such long gangs could not have been used. Lumber was purchased at Rossport and fashioned into fish boxes.

In winter, many miles of net needed repair. Also in winter, the Nicolls put up ice from frozen Coldwell Harbour, for later use in summer, to pack fish and boxes for shipment on the Canadian Pacific. In good seasons—the best were in the 1920s and 1930s—average daily yields were one and a half tons, and seasonal catches from this one fishery alone averaged close to a half-million pounds. Deaths of the owners coincided with the collapse of the fishery owing to the damage caused by the sea lamprey, and "Nicoll Bros., Producer of Fish," closed down in 1956. Today, Port Coldwell remains a ghost fishing village, with decaying and charred docks, buildings and foundations, and hulls of fishing vessels.

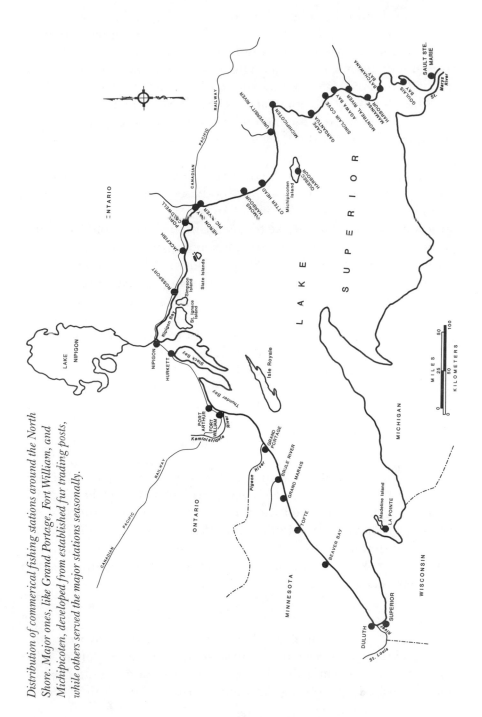

*Distribution of commerical fishing stations around the North Shore. Major ones, like Grand Portage, Fort William, and Michipicoten, developed from established fur trading posts, while others served the major stations seasonally.*

187

Even before the days of government regulation of assigned fishing areas, fishermen arranged their own territorial boundaries so as not to overlap, although occasional disputes arose, with consequent damage to gear. The system operated largely with major *fisherman-shippers* located at Fort William and Port Arthur on Thunder Bay, Nipigon Bay, and Sault Ste. Marie; these shippers bought fish from the men who did the netting and catching, and they distributed the harvest by ship and rail to many distant markets.

Early Canadian fishing comprised mainly the operation of Booth Fisheries Company, a Chicago-based "fish trust" made up of about ten companies, from Detroit to Manitoba. Booth's operation was extended into Lake Superior in the 1890s, and the company picked up fish at many stations along the eastern shore for shipping to Sault Ste. Marie. Stations included several along the Pukaskwa Peninsula—Simons Harbour, Dog (University) River— as well as Quebec Harbour on Michipicoten Island and Gargantua Harbour.

The Nipigon Bay Fish Company flourished in Rossport as a major competitor to Booth Fisheries and survived to 1953, when the lake trout population crashed.

Booth's operations were later taken over by James Purvis and Son, Ltd. In 1959 Purvis sold to Ferroclad Fisheries, located at Mamainse Harbour, and the company is still operating today.

In the Thunder Bay-Black Bay area, the first private company was the Lake Superior Fish Company, established in 1879, which shipped mainly to Minneapolis. This company in 1884 became the Port Arthur Fish Company, which then supplied Booth Fisheries. Others included the Union Fish Company and Gagne Brothers Fishery in the 1890s, and the Fort William Fish Company, founded in 1916, and, at Hurkett, Nuttall Fisheries in 1919. Later, Kemp Fisheries became the dominant fishing force in the Thunder Bay area and is still operating.

The lake whitefish was the most favored and heavily exploited species in the first years of Lake Superior fishing. Early accounts attest to the delicate flavor, and appetites were not diminished by even regular daily consumption. Complementing the desirability of whitefish on the table was the relative ease with which these fish could be captured.

Because the whitefish occupied shallow, nearshore waters during its entire life history, its population could be heavily exploited with the primitive vessels and fishing gear of the times. In fact, in the first year for which reliable statistics on commercial catch were recorded, 1885, the annual catch of lake whitefish exceeded five million pounds in the whole of Lake Superior—an all-time high that was never matched.

With its exceptional favor as food and the ease with which it could be taken, little wonder that the annual catch of Lake Superior whitefish declined through overfishing in the first few decades of modern commercial fishing industry to only about a half-million pounds. Contributing also to this decline apparently was the effect of sawdust, bark, and other lumbering wastes, suffocating whitefish eggs on spawning areas.

During the late 1950s predation by sea lampreys effected a drastic decline to an all-time low of a little more than one-third million pounds.

After effective lamprey control was initiated in the 1960s, whitefish catches increased once again, to about one million pounds. Although fairly stable at a level of one million to one and a half million pounds in all of Lake Superior for the past several years, the lake whitefish does not now support a major commercial fishery.

Along the North Shore, whitefish catch overall has been by far the greatest in Canadian waters, where there is a greater occurrence of appropriate shallow-water habitats, in contrast to the deep waters along the steep Minnesota coast. Actually, Ontario has reported whitefish landings since 1867, catches that, with a few exceptions, have been remarkably stable at about 200,000 to 400,000 pounds per year. The low year on the North Shore—less than 100,000 pounds, all from Ontario—was reached in 1960 after the sea lamprey invasion; recovery owing to lamprey control has resulted in increased catches to previous levels.

With the decline of whitefish in Lake Superior and improved boats and fishing gear, the lake trout soon became the leading commercial species. But in contrast to the whitefish, lake trout were harvested at relatively stable levels—right from 1885 to the edge of the decline brought about by the sea lamprey—fluctuating slightly around four million pounds per year, for all of Lake Superior. North Shore waters yielded about half, or two million pounds per year.

Several reasons to explain the difference in trends between whitefish and trout have been put forth. The main one appears to be that the lake trout, occurring in many quasi-discrete stocks located in shallow as well as deep environments, near isolated offshore reefs and islands, even in tributary streams, were subjected to a "fishing-up" process in which the stocks were sequentially located and exploited. The fishing-up theory also applied to sizes of lake trout caught: as fishing efforts were first directed toward the largest fish, and their numbers declined, increasing effort was applied to smaller trout.

These two factors, resulting in the maintenance of high yields, gave the appearance of a stable population, when in fact the overall lake trout population was declining unseen.

Furthermore, with primitive methods and equipment in the early years, lake trout could be fished only during seasonal migrations to shallow water, in the fall, whereas whitefish could be harvested in nearshore waters virtually year-round.

Some abundance indicators suggest that actual Lake Superior trout populations were declining slightly during the period of intensive fishing, 1885-1953. The lamprey then administered the *coup de grâce* to the lake trout. Catches plummeted by an alarming rate, to a low in 1961 of about one-third million pounds for Lake Superior, only 50,000 pounds on the North Shore, way down from two million.

Since initiation of rehabilitation programs of lamprey control and trout stocking in 1962, commercial fishing for lake trout has been greatly curtailed in all of Lake Superior, except for assessment purposes. Assessment catches have thus been kept artificially low at about one-fourth to one-half million pounds per year, only one-fifth million pounds on the North Shore.

Abundance indicators, however, suggest increased populations, as measures of success of the lamprey control and stocking programs. Hopes for the return of the delectable lake trout to the fish market and supper table remain high, albeit cautious.

The commercial fisheries for herring, chubs, and smelt were much slower to start up than those for whitefish and lake trout—understandably, in view of the greater economic return from the two larger species. Beginning in the first two decades of the 1900s, catches of herring from the lake as a whole increased rapidly, owing to greater effort applied to this species as whitefish declined. Total yield from Lake Superior rapidly rose to about ten million pounds per year about 1915, and for many of the next fifty years it ranged between ten and fifteen million pounds, even approaching twenty million in the peak year, 1941. The greatest proportion of this catch was from United States waters, particularly along the south shore, although Minnesota's portion of the North Shore herring production also was high. In 1927, more than nine million pounds of herring were taken from Minnesota waters. Herring catches from Ontario's portion were low but fairly stable during this period, at around one to two million pounds.

But declines set in about the mid-1930s, from which the herring has not fully recovered. Scientists disagree about the reasons for the decline in the past fifty years. Competition for the food resource, plankton—which the herring shared with an increased population of small chubs and, especially, with the introduced smelt—may be one reason. Another possibility is predation by smelt upon larval herring, a factor that probably became much more important when the population of exotic smelt increased after arrival of the species in Lake Superior in 1930. Overfishing, also, is heavily indicted; in Minnesota waters, for example, the commercial catch in the heyday of the 1920s and 1930s amounted to more than four pounds per acre of lake waters fished. Biologists now calculate that the infertile waters of Lake Superior should yield only about one pound per acre for a sustainable catch.

With the loss of lake trout and whitefish owing to the sea lamprey in the 1950s, herring remained as the only abundant marketable species. Lamprey predation on large herring, and selective fishing effort directed toward the herring, were other factors probably contributing to the decline.

Today, herring catches remain at about three million pounds per year for all of Lake Superior, with Ontario yields, particularly in the Thunder Bay-Black Bay area, constituting the bulk of the

Lake Superior commercial landings. This Canadian herring fishery on the North Shore remains today as the largest commercial fishery for any species in Lake Superior.

The chubs, with their deepwater habit, were fished with fine-meshed gill nets set on the bottom. Chubs were used primarily for smoking; they did not contribute a large fresh fish market. Originally the group consisted of five species in Lake Superior, but commencing with the largest of the five, fishing effort was sequentially directed toward the next smaller species. All but one species of chub were soon eliminated as a commercially valuable fish.

The chubs today have hybridized into a complex population, in which separate species are almost impossible to identify. Only the bloater—the smallest of the five—remains as a fishable population, and there is evidence that today's bloater is a complex hybrid, significantly larger than the original. The sea lamprey obviously had no direct effect upon the chubs, for their decline to only the one species occurred before the lamprey arrived. Modern catches of chubs range from one to two million pounds per year, with the North Shore contributing about 20 to 40 percent, and most of that from Ontario waters.

The smelt is the only exotic species entering into Lake Superior's commercial fishery. First taken incidentally in herring nets in 1946, smelt entered the regular fishery in 1952. The whole-lake catch increased rapidly between 1952 and 1961, to nearly one and a half million pounds per year. In the years 1958-60, smelt were so abundant that they fouled gill nets, of any size, set for other species and were considered a great nuisance by many commercial fishermen. After that, catches then leveled off between one and two million pounds per year for two decades, although a peak of four million pounds was recorded in 1976. The bulk of the smelt harvest—around 75 percent—has been from Minnesota's North Shore waters. Large die-offs occurred, however, in the early 1980s—reason unknown—with declines in catch to about a half-million pounds.

Commercial catches, like fish populations in the lake, have followed a complex trail. Changes in fishing efforts, gear development, and effects of exotic introductions have all affected the commercial harvest in complicated ways. And like Lake Superior's fish populations, too, commercial yields are still changing and dynamic.

The commercial catch of all species from Lake Superior has ranged from about a half-million pounds per year in the first years of systematic commercial harvest, about 1870, to a high of twenty-five million in 1941. The 1930s and 1940s were the halcyon years, with total annual catches, spurred by food needs in the Second World War, reaching around twenty million pounds. Fishing effort was intense in those years—too intense, we know now—and times were good, or so it seemed. The sea lamprey's arrival was the factor that assaulted fish populations already strained to their limit, and the fishing industry essentially collapsed.

In recent years total catches for all species have been less than ten million pounds per year. Even this figure, low as it appears compared to previous highs, masks the overall effect of the fishery collapse, for the annual herring catch in the Lakehead area alone comprises about two to three million pounds, or about one-third of the total Lake Superior commercial catch of all species.

Early entrepreneurial activities in the North Shore area stimulated exploitation of the region's natural resources: fur, minerals, timber. Fish played an important part, providing food for traders, prospectors and miners, and lumbermen, but commercial trade in fish was limited to local distribution. Expansion awaited the development of long-range transportation to distant markets, which began in the late 1800s and then expanded greatly in the first half of the 1900s.

This half-century was a period of extremely heavy fishing, in both Ontario and Minnesota. Huge catches were spurred by improved equipment and techniques, rapid transportation, and distant markets clamoring for the delectable products of Lake Superior. Actually, a gradual decline had set in as early as 1915, but few noticed, and harvests remained high through increased effort, setting the stage for the disaster to follow.

Now commercial fishing has declined to near insignificance along many parts of the shore. Along the more extensive Ontario shore, the number of commercial licenses is down to only about eighty, all with quotas on lake trout. In Minnesota, only nine fishermen are licensed for lake trout, strictly limited in quotas and seasons.

But the fish resource itself—now with different species, new management objectives, innovative techniques for harvesting— may again be the base of new economic industry. Sport angling, with all its attendant relationships to tourism, resource appreciation, and recreation, will, under proper management, utilize this resource of incalculable aesthetic and economic value in the future—to become perhaps the greatest of all North Shore industry.

# TWELVE

# *Fishing for Fun*

The man in the aluminum canoe feathered his paddle, rotated it slightly, and the light craft curved around a small point of land. At the very tip of the point, a small spruce at first obscured the man's vision, but as the canoe glided around farther he could see the little bay, a curving indentation in the hedge of leatherleaf and sweet gale that lined the shore. Upon the surface of the otherwise quiet water, ripples caused by a rising fish spread and faded. The man let the canoe drift.

Quietly he picked up a fly rod. Stripping a few loops of line, he worked out a cast of perhaps fifty feet, then lightly let a small squirrel-tail streamer fall about a foot from the leatherleaf. The fly sank, and the man waited. Judging the streamer to be well under the surface, he began a slow, jerky retrieve.

Almost immediately a boil of water rose near the end of the man's leader. Then the boil erupted in a sudden crash of foam and amber water. The man reacted instantly with his strike, and the fly rod arced and throbbed. In the center of the exploding white water had been the unmistakable flash of the bright red flanks of a large trout.

Five minutes later, after the fish had pulled the light canoe around the little bay, the man leaned over the gunwale and netted

*Speckled Trout Creek*

the bright fish. It lay on the canoe deck, spent but still vital, its gill covers moving slowly, its spotted sides and vermilion belly flanks wet and glinting. The man sat quietly, resting, admiring the brilliant colors.

The fish was a splake—a hybrid between a brook, or speckled, trout and a lake trout. Marking and coloration showed the splake's inheritance from both parents: irregular-shaped, light spotting on the gray-green back, but not as gray as the lake trout's; bright and round spots along the sides, but not as bright or as red as the brook trout's; on both sides of the belly, the flanks reddened from the belly upward, the prespawning coloration of the male brook trout. The splake weighed a good two pounds and measured about seventeen inches.

The time: mid-September. The place: a small remote lake, inland from the North Shore of Lake Superior.

The fishing experience that this canoeman had just enjoyed was not at all representative of pristine fishing in a northern wilderness. Neither the circumstances in the small lake nor the fish itself was natural, but rather the product of modern fishery management.

The little lake probably had first held a fish population of stunted yellow perch or suckers, or perhaps no fish at all. Fisheries management personnel, after determining that the lake was suitable for other fishes, in particular for trout, had reclaimed the lake—that is, treated it with toxicants to remove all existing fish. Restocking with hatchery-reared trout had established a new fishery.

The splake is also a product of human technology—the artificial hybridization of two natural species: the offspring of female lake trout and male brook trout. Developed for stocking in the Great Lakes, primarily in Lake Huron as a replacement for the lake trout, the splake combines the large size of lake trout with the early maturity of brook trout. Furthermore, the splake grows faster than either of its two parents. Its use in small, inland lakes provides the opportunity for an unusual sport fishery, adding to the diversity of recreational resources.

The splake is an aggressive feeder and, like both its parents, is readily taken by angling. Consequently, to keep fishing pres-

197

sure light and avoid rapid exploitation, and also to retain an atmosphere of quiet and natural beauty, the access to these lakes is kept primitive, usually by at least a short canoe portage.

The planting of splake trout in these small inland lakes is relatively rare, and more commonly brook trout and rainbow trout are stocked. Like all our salmonids except lake trout, they require the flowing currents and gravel riffles of streams for spawning. Like the splake, brook and rainbow trout must be continually restocked with hatchery-reared fish in order to maintain these inland lake fisheries.

The choice of species depends upon environmental conditions. The brook trout is usually the preferred choice, it is native to the Lake Superior drainage and considered by many anglers to be the most colorful and beautiful of all trout. Of the two, the brook trout, however, is the most demanding in its requirement for cool water. The rainbow is frequently stocked in lakes with slightly warmer waters and grows to a larger size. In some respects, the rainbow is a greater challenge to the angler, is caught less easily, and thus survives better to provide good fishing later in the season, as well as in later years. The rainbow has been more successful than the brook in larger, more fertile lakes. Brown trout are less frequently used in lakes, owing to their slow growth and great resistance to angling.

In a few cases, some unusual fishes such as Arctic grayling, kokanee (landlocked sockeye salmon), coho salmon, and Atlantic salmon have been tried, mostly to develop novelty fisheries.

In most cases, only a single species of trout is stocked, while in others brook and rainbow are combined. Other competing fishes are kept out. Fishing with live minnows for bait is prohibited, for obvious reasons.

The selection of small inland lakes for reclamation and establishment of stream trout fisheries is a rigorous procedure. Lake morphology, water clarity, and surrounding forest together contribute toward producing the right combination of water temperature and dissolved oxygen, criteria that are particularly demanding for trout. Usually selected are seepage lakes, that is, lakes without inlets or outlets, to avoid reintroduction of unwanted fish species. Lakes so small they may not even be counted as lakes—five to ten acres, or even smaller—are often selected;

though small in surface area, these frequently are deep, maybe fifty feet.

Bog lakes, with soft, infertile water that assists in keeping oxygen levels up, are sometimes selected. The small bog lake with its shores lined with a leatherleaf-*Sphagnum* mat, including cranberries, Labrador tea, pitcher plants, and a variety of orchids, produces a wild, northern atmosphere. On a cool, quiet evening, the fragrance of sweet gale, crash of a beaver tail-slap, or cry of a loon may appropriately complement the splash of a rising trout.

Suitable lakes occur in both the Minnesota and Ontario portions of the North Shore, where lake reclamation with toxicants is commonly performed. Minnesota lists about sixty of such lakes. Many small lakes not needing reclamation occur in the Ontario portion, yet these also require regular stocking. Some of the best of these are on wild Black Bay Peninsula or on the larger offshore islands, such as St. Ignace Island.

The brook trout (*Salvelinus fontinalis*) was native to Lake Superior and inhabited all streams tributary to the northern coast. Along with the lake trout, it was the preferred sport fish of early inhabitants on the North Shore. Spawning in the fall, the brook trout leaves its eggs to incubate through the winter beneath the coarse gravel of its redd, or spawning bed. During incubation, a continuous supply of dissolved oxygen is required in water flowing through the gravel.

Hatching takes place in early spring, and after the tiny fry work their way upward through the redd's gravel, they are on their own in a watery world of stream currents, bottom stones and gravel, sunken logs and stumps. Among these substrates, the brook trout spends most of its relatively short life searching out food: small aquatic invertebrate animals, mostly insects, such as immature mayflies, caddisflies, and midges. Fortunately for the fly rod-wielding angler, the brook trout also feeds readily on insects floating on the stream surface—both aquatic insects newly emerging as adults and terrestrial forms that drop onto the water—which the angler imitates with artificial dry flies.

The life cycle of the brook trout in most small streams is relatively short: three to four years. Sexual maturity comes early, and most brook trout spawn in the fall of their second year of life, some males even in their first year. Given all its requirements in its stream world—food, cover against predators, sustained cool water and oxygen, and good spawning gravel—brook trout populations will live and prosper indefinitely in a short reach of stream of only 100 yards or so.

Brook trout in their usual small stream do not reach great sizes. Typical of most angler catches from North Shore streams would be fish less than ten inches long, and the occasional lunker of twelve inches would be a rare five- or six-year-old.

The brook trout occurs in some exceptional circumstances. For example, some migratory populations that occur in Ontario hatch in stream redds but migrate downstream to live most of their lives in Lake Superior, returning later as adults to spawn. In the lake, the brook trout is known as a *coaster*, reaching a much larger size than its stream cohort. The coaster, however, has a much different appearance than its stream cousin, lacking the usual bright coloration and appearing more like a lake trout. It is taken rarely by the casual lakeshore angler, but some have developed the unique fishing skills needed to find and catch the coaster. Many are taken incidentally by anglers trolling for lake trout.

Some of the best coaster fishing is in the Nipigon Bay area, in Black Bay, and around the offshore islands. Interestingly, these brook trout did not suffer the same declines as a result of lamprey depredation that the lake trout experienced, no doubt owing to their smaller size, early maturity, and inland stream spawning. Ontario lists seventeen streams used by spawning coasters; although once common in Minnesota waters, this anadromous brook trout is now rare in the state.

A second example of an atypical brook trout population is in the beaver pond. Damming up small streams to facilitate food gathering and home building, the beaver creates an *impoundment* in a brook trout stream—and thus adds a totally new dimension to the fish's habitat.

Beaver ponds on trout streams are controversial. If the stream is originally cold, as most small North Shore streams are, the warmed waters of a pond may be beneficial to fish growth. Furthermore, the added area of a pond may increase greatly the production of aquatic invertebrates used by fish as food. Perhaps most important to the angler, however, is that the increased space and depth, added to a small stream with tiny habitat niches, may provide for substantially larger specimens—perhaps some of the five-year-olds.

On the other hand, beaver dams on a stream already thermally marginal can cause the stream to become too warm for trout, as well as destroy insect-producing riffles.

All beaver ponds are impermanent, requiring constant maintenance by the beaver. Ponds may exist for only a few years, may be replaced in a different location, or may be rebuilt years later after draining or being damaged in a flood. Many are built in remote locations, and this lack of accessibility, combined with their transitory nature, means that fishing success will be uncertain.

Even for the intrepid and skilled bush-whacking brook trout angler, searching for a new beaver pond and large trout becomes a game of alluring adventure, hard work and many mosquito bites, and occasional success.

By the time brook trout arrived in Lake Superior a few thousand years ago, water elevations following glaciation had already fallen to at least near modern levels. All around the North Shore, the high, rugged coastline caused tributary streams to plunge in falls and cataracts to the big lake. Consequently, the brook trout was unable to ascend and colonize the streams very far. And when early white traders and prospectors tried their hand at angling in the many headwater streams that should have been trout streams, they found the streams barren. But the stream environments were highly suitable for trout, and later fish culturists and biologists introduced the brook trout above barriers in many streams and thus to hundreds of miles of headwater streams around the North Shore.

Along the Minnesota coast, stream systems are simple and short, and brook trout have now penetrated the most remote headwater trickles. Inland from Ontario's Lakehead region and the eastern shore, however, stream networks are much larger and more complex; and whereas almost all streams near Lake Superior are inhabited by brook trout, not all headwater reaches above remote waterfalls have received the brook trout. Perhaps they will, eventually.

One major exception to the common pattern of brook trout distribution is Lake Nipigon. At some time during the retreat of

201

glaciation, perhaps because the Nipigon Basin received glacial meltwaters differently than did the Superior Basin, brook trout apparently became established in Lake Nipigon separately from Lake Superior. After glacial lake levels had receded, major waterfalls on the Nipigon River prevented any dispersal of Lake Superior fish to the Nipigon, and this upper brook trout population became reproductively isolated.

Nipigon brook trout, now perhaps genetically distinct, have a high growth rate, and trout three to four years old are trophies indeed. With the capture in 1915 in the Nipigon River of the world's record brook trout—fourteen and one-half pounds—Nipigon brook trout fishing became legendary. The purity of the strain is maintained by Ontario's fishery managers, and fishing today continues to embrace many beautiful, trophy-sized brook trout.

The lake trout (*Salvelinus namaycush*) is a close relative of the brook trout, but it varies in several respects important to the sport angler. First, the lake trout grows to a much larger size, and second, it is a lake-living fish that is angled extensively in Lake Superior and many inland lakes.

Another important difference in the biology of these two species is that whereas the brook trout must have the flowing water and stream-bottom gravels for spawning, the lake trout spawns in lakes. It is the only species of North American trout and salmon to do so. The natural distribution of the lake trout also differs from that of brook trout; the lake trout occurs all across northern North America, approximating the same distribution as Pleistocene glaciation. No doubt the lake trout's ability to spawn in lakes, rather than being limited to streams, facilitated this distribution following glaciation. Consequently, we find the lake trout occurring naturally in deep, cold lakes inland from the North Shore, in addition to Lake Superior itself.

Spawning by lake trout occurs over rocky bottoms with boulders and cobbles, in relatively shallow water. No redd is constructed, but the lake trout often does some processing of the bottom in cleaning rocks by fanning with its body and fins, and a certain amount of rubbing with its snout. Eggs and sperm are ejected together by the two parents, and the fertilized eggs fall to the rocky bottom and settle among the rocky crevices. Thus protected, the eggs incubate through the winter, for four to five months, and hatch in spring. Lake trout fry, like brook trout fry, struggle out of their rocky womb to make their own way and search for small items of food—small aquatic invertebrates—among the rocks and gravel of lake shoals.

Like the brook trout, the lake trout must have cold water. In inland lakes, the lake trout may be available to the angler in shallow water in early spring as the ice breaks up, whereas later in summer deepwater trolling may be necessary. But again, in the fall, as the time for spawning approaches and surface waters cool, angling in nearshore shallows may produce most lake trout. Ice fishing in inland lakes can be highly successful, too, and with the advent of the snowmobile for access to roadless areas, fishing has been so good that overfishing in some lakes has occurred.

Around the entire North Shore, nearly 300 lakes occur with known lake trout populations, almost two-thirds of these in Ontario and one-third in Minnesota.

Lake Superior provides the greatest source of lake trout angling, essentially year-round. Cold water, even at the surface in summer, allows lake trout fishing at almost all times. Movement of most trout to nearshore waters in the fall produces the best angling success at that time. Some closed seasons are in force.

Since fall spawning occurs over rocky bottoms in shoal areas, some of the best Lake Superior lake trout fishing is around offshore reefs and islands. Isle Royale, the Slate Islands, and the many offshore islands in Nipigon Bay are all known for their excellent lake trout angling. Winter fishing through ice is increasingly popular in sheltered areas, for example, Thunder Bay.

Lake trout in the anglers' catch weigh down the creel much more than do the diminutive brook trout. Most hook-and-line-caught lake trout average two to four pounds, less commonly up to ten pounds, even more. Larger trophies are possible, however; the North American angling record came from Lake Superior in 1952: sixty-three pounds, two ounces. The largest lake trout taken by any method, however, was a whopping 102 pounds, taken by gill net in Lake Athabasca in 1961.

Lake trout, like all trout and salmon, are predators, feeding on other animals. In early life history stages, their diet consists of small invertebrates: insects, aquatic worms, crustaceans. In later years, the lake trout is largely piscivorous, feeding on other fish. Consequently, most angling is done by trolling or casting lures that imitate a small fish—such as spoons and fishlike lures—or by fishing with live bait like smelt and other small fishes. Probably most angling is done from boats, but shore fishing with lures or bait is becoming increasingly popular in late summer and fall, when lake trout approach shallow spawning grounds. Much of Lake Superior's shoreline is too deep and rocky for wading and fishing, but casting from stream mouths, where the extension of stream currents out into the lake can help get lures out to deeper water, may be especially effective.

In Lake Superior, angling as well as commercial fishing for lake trout declined drastically after the invasion of the sea lamprey. In recent years, lake trout populations have been on the rise, owing to lamprey control and stocking programs, and angling success has likewise been on the increase. The lake trout, like its stream cousin the brook trout, is particularly susceptible to the hook. And whereas commercial fishing with nets and traps can be tightly controlled, it is much more difficult to regulate the sport fishery, or even to measure its catch. In an overall plan to restore the lake trout, it may be necessary to adjust the sport harvest more tightly.

The lake trout is a fragile species. Beset on many sides—a potentially intense commercial fishery, critical environmental requirements, predation by the sea lamprey, competition from introduced species, and a booming recreational fishery—it is in a precarious position. The lake trout will require some intensive management if we are to preserve high-quality sport fishing for this favored native of the North.

The superb sporting qualities of the rainbow trout (*Salmo gairdneri*) were early recognized by anglers on the Pacific coast. Anadromous rainbows ascended coastal streams on annual spawning runs and were taken by fishermen who were thrilled by the long, heavy runs and high leaps of these strong fish. The term "rainbow" probably comes from the pink or red stripe down the fish's flanks. Otherwise, the fish is silvery overall, with perhaps a glint of steel-gray on its head and back. With great respect, the name "steelhead" was given to these river-running rainbows.

In the latter half of the nineteenth century, development of fish culture and fish hatcheries was well under way in North America. Around 1875, fish culturists, acting privately out of their personal interests in fishing and natural history, began to experiment with hatching, rearing, and transplanting the rainbow trout to eastern North America. Streams in New York, Pennsylvania, Michigan, and Wisconsin were soon receiving plantings of rainbows. And, in 1883, the provincial government of Ontario introduced the rainbow trout to Lake Superior with some fish planted near Sault Ste. Marie. Other plantings were made in later years, 1889 and the 1890s, in Lake Superior tributaries in Michigan, Wisconsin, and Minnesota, as well as on Isle Royale. By the end of the century, more than one million rainbow trout had been stocked in the Lake Superior basin, many of these along the North Shore.

The transplanted rainbow was immediately successful. Rapidly colonizing all suitable spawning streams, this species apparently found an ecological niche unoccupied by other species. The first capture of a rainbow in Lake Superior was by a commercial net along the North Shore in 1895; sport anglers were taking fish in North Shore streams by 1897. By 1900, rainbows were taken in commercial fishing nets all around the North Shore, and sport fishermen were taking thousands of adult spawning steelhead from North Shore streams.

The rainbow gained in popularity through the first half of the twentieth century. Many will say that it is the fightingest fish in fresh water—an accolade stemming from its habit of high, strong leaps on the end of a line. The surprising strength of a fighting steelhead, fresh and bright from Lake Superior, is sometimes hard to believe. Such an experience—more often than not ending with a parted leader and lost fish—can leave the angler with trembling knees and shattered nerves, not to mention a lifelong addiction to steelheading. Maybe the "rainbow" derives from a similarity to the atmospheric kind, while at the apogee of a high leap over a rushing river rapids.

When the sea lamprey invaded the Great Lakes, the rainbow, like the lake trout, fell prey to this marine predator. Numbers declined drastically. With the implementation of lamprey control and further rainbow trout stocking, the steelhead now has made a remarkable recovery.

Today, the steelhead ascends virtually every suitable spawning stream along the North Shore to build its redds and leave its fertilized eggs. Unlike the brook and lake trout, which spawn in the fall, the rainbow is a spring spawner and runs when the first warm rains muddy up tributary streams, usually about mid-April to mid-May.

Rainbow eggs incubate in their rocky redds for several weeks during the spring and hatch into tiny fry in the warmer, more hospitable days of late spring. The juvenile does not immediately run to Lake Superior, however, but remains in its natal stream usually for two years. During this time, growth is slow. At about six to eight inches in length, the juveniles *smolt*—that is, undergo physiological and color changes—and move down to Lake Superior, just as their forebears did to the Pacific Ocean. In the big lake, growth is accelerated, and after another two or three years in the lake, the steelhead reaches a weight of about four to six pounds, matures sexually, and is ready to return to a stream for spawning. Homing to the natal stream is common but not invariable.

It is, of course, on these spring spawning runs that the steelhead becomes available to the angler. The harvest during spawning does not appear to reduce the steelhead's reproductive capacity, since fecundity is sufficiently high—at several thousand eggs per female—to compensate for angling losses.

Unlike the Pacific salmon, the rainbow does not die after spawning but returns to Lake Superior. It has the capability to spawn again in a future year, although few survive to do so.

During the spring spawning run, fish feed readily on eggs of other rainbows, commonly present at this time. Consequently, baits of natural spawn bags or eggs, or imitations of eggs, are preferred. At this time of year, waters are cold and weather unpredictably inclement; steelhead anglers must be a tough breed.

The rainbow also runs in late fall, in addition to the spring run. We're not entirely sure why this happens—some inheritance of Pacific coast spawning behavior, perhaps. Some anglers swear that fall fishing is the better. Fish are brighter, fresher from the big lake, less enervated from the production of eggs and sperm, and more vigorous in their fighting displays. Fall fishing, mostly in November with advancing cold weather, autumn storms, and rain-swollen rivers, also requires an unusually sturdy breed of angler.

Minnesota has experimented with a fall-spawning rainbow—the Kamloops strain—but the progeny from these apparently soon revert to the more normal spring-spawning habit.

With anglers capturing most fish on their first spawning journey, the average size runs about five pounds; repeat spawners weigh in heavier, with a weight of over twelve pounds not really uncommon. The record rainbow from North Shore waters weighed seventeen pounds, six ounces, and was taken from the Knife River, Minnesota, in 1974.

The rainbow is a great leaper—over waterfalls as well as on a line—and will ascend its spawning stream as far as it can. However, there are limits to its leaping ability, and the tumultuous character of most North Shore streams all around the coast prevent the rainbow's ascent very far. The stream reach between the mouth in Lake Superior and the first insuperable barrier of waterfalls is sometimes very short, often less than one mile. The nursery area for juvenile steelhead may thus be quite small. Since the juveniles must have all their needs for food and cover met in the stream for their first two years, it is clear that the factors limiting juvenile rainbow production are related to the quality and quantity of stream resources, rather than to the number of spawning adults.

In the original stream conditions of presettlement years, it was the brook trout that inhabited lower stream reaches. Today, the lower reaches are utilized mainly as nursery areas by young steelhead, and the upper reaches, above barrier waterfalls, by brook trout.

The lower reaches, too, are the scene of angling and much fighting display by adult steelheads, during spawning runs of Lake Superior-grown rainbows.

With spawning in virtually every suitable stream around the coast, large and small, a great resource of streams is available for steelhead angling. Ontario lists over 100 streams with known rainbow spawning, with thirty of these receiving major runs. Among the best known are the Kaministiquia and Nipigon at the Lakehead, Jackpine, Cypress, and Steel rivers along the northern

207

shore; the White, Pukaskwa, and University rivers in the Pukaskwa Peninsula; and the Michipicoten, Agawa, and Montreal rivers on the eastern shore. Minnesota claims about sixty streams accessible to spawning, and about twenty of these with major runs; most popular are the Knife, Baptism, Cross, Devil Track, and Brule rivers. The steelhead is also attracted to other, smaller creeks and tiny tributaries, where angling, even if impossible in the brushy little rivulets, may nevertheless be rewarding near stream mouths in Lake Superior. Willingness to explore off the beaten trails is often the key to angling success.


Inland from the coast lie many hundreds of lakes, glacier-scoured into Precambrian bedrock. Most interconnect by short streams or small flowages to form a vast network of waterways, largely navigable. Some are connected by major rivers like Minnesota's Brule, the border Pigeon, and Ontario's Sand River on the eastern shore, in which the lakes form an integral part of modern recreational canoe routes.

But these inland lakes and streams also provide a recreational fishery that, in terms of numbers in the catch, is unequaled in the North Shore region—a fishery for the walleye and northern pike. The catch of these two species constitutes at least three-fourths of the total sport catch in the land around the North Shore.

The walleye (*Stizostedion vitreum*) is classified in a group we term *coolwater* fishes—that is, preferring waters not as cold as trout waters but still not as warm as those inhabited by sunfish and bass. The walleye's spawning requirements include rocky shorelines, similar to the lake trout's. Consequently, the walleye found a ready home in the myriad cool, deep, rock-lined lakes of the Precambrian Shield and occurred there as a native fish. It is frequently found in lake trout lakes, where the trout inhabits the colder depths.

Like the lake trout, too, the walleye is a piscivore. Occurring commonly in walleye lakes north of Lake Superior are the walleye's principal forage fishes, the tullibee, an inland lake form of the lake herring, and the yellow perch, both also coolwater species. As a favored table fish, the walleye fed many an early trav-

*Location of some important sport fisheries along the shore, near islands, in bays, and on inland streams and lakes.*

eler and trader in the Lake Superior country, as it still does today for modern backcountry canoeists and campers.

The northern pike (*Esox lucius*) is sometimes also classed as a coolwater fish, since it inhabits north temperate waters and is commonly found along with the walleye and yellow perch. However, the northern is much more than that. The northern could almost be termed cosmopolitan, so widespread is its distribution in many kinds of waters, including a circumpolar distribution in the northern parts of both eastern and western hemispheres.

Northern pike require marshes, ponds, or shallow backwaters for spawning, and if these backwater areas occur along lakeshores or tributary to a good walleye lake—and they usually do, in the North Shore region—the angler is likely to find both walleye and northern pike on the same fishing trip. The northern is also a piscivore—a particularly ravenous one, in fact—and like the walleye also forages on yellow perch, as well as on other small fish, including smaller northern pike. Lures used in angling commonly imitate small fish, for both species. Because of the difference in spawning habits, however, the more common fish taken in rivers and streams is likely to be the northern pike.

Both the walleye and northern pike grow to large sizes, and specimens of eight or ten pounds are not rarities. Northerns over twenty pounds happen along with some regularity. Both are long-lived fishes, with individuals ten years old fairly common; consequently, neither species can withstand especially heavy fishing pressure and still provide large trophies.

The inland lake region of the North Shore is well suited to management of these two fishes. A great abundance of lakes, with varying access, disperses angling pressure. In lakes easy of access, it will be a bit tougher to creel a limit. But the more remote waters, difficult of access and set in the wild solitude of northern forests, will provide the best chance of larger catches— as well as catches of larger walleyes and northerns.

In some of the cool inland lakes proliferates *Micropterus dolomieui*—the smallmouth bass. A hard-fighting, acrobatic sporting fish, the smallmouth provides high-quality angling. Smallmouths occur mainly in lakes not far inland from the coast. Lakes in the Thunder Bay area, especially in the region directly west, just north of the Pigeon River, are notable. Not as common as the walleye and northern pike, it is uncertain whether the smallmouth was native to the North Shore drainage.

Another species in the sport fisheries of inland lakes is the lake whitefish (*Coregonus clupeaformis*), the same whitefish species so important in the commercial fisheries of Lake Superior. The whitefish is widely distributed in inland Ontario lakes, where it is caught with small spinners or minnows for bait. Unlike many other fishes, the whitefish continues to feed actively in winter, and thus it makes prime fishing through the ice. Broiled, fresh whitefish is delectable eating.

Three new fishes, strangers from the Pacific coast, now figure prominently in the Lake Superior sport anglers' catch: the coho, chinook, and pink salmon. The successful establishment of the first two, coho and chinook, was the result of deliberate introduction of these exotic species to revitalize the Great Lakes sport fishery after the invasion of the sea lamprey and the decline of the lake trout. The introduction of pink salmon was an accident on the North Shore of Lake Superior, resulting in its totally unpredictable success throughout the Great Lakes.

In the wake of the upheaval caused by the sea lamprey, Lakes Huron and Michigan were invaded by the marine alewife, a small silvery fish in the ocean herring family, with high reproductive potential. Like the lamprey, the alewife apparently dispersed from Lake Ontario to Lake Erie and the upper Great Lakes through the Welland Canal. With suitable environmental conditions in Lakes Huron and Michigan, and in the absence of a natural predator, the alewife increased to enormous numbers.

The coho and chinook salmon, it was thought, not only would help control alewives by using them as forage, but in so doing would also survive well and grow to large sizes, providing an important sport fishery. The salmon introductions were to be a huge success.

In 1966, when attainment of sea lamprey control appeared assured, the state of Michigan made the initial salmon plantings in the Great Lakes: 660,000 coho salmon in Lake Michigan and 192,000 in Lake Superior. Later plantings of coho along the North Shore of Lake Superior were made by Ontario (1969-71) and Minnesota (1969-74).

The coho, or silver, salmon (*Oncorhynchus kisutch*) is a favorite sport fish in the Pacific Northwest. It ascends coastal tributary streams for long distances and is caught by stream anglers far inland from the ocean.

The coho's initial planting in Lake Michigan was immediately successful, and soon large specimens of fifteen to twenty pounds were frequently taken.

But the coho is a fall spawner, in late October and November, and on the North Shore it failed to provide good fishing. The size of coho in Lake Superior was much smaller than in Lake Michigan, two to six pounds. Unsuccessful in spawning at first, the coho today is reproducing in eastern streams along the Ontario North Shore and progressing westward. Some winter fishing along the shore, especially near stream mouths, has been successful, but fish were often badly deteriorated. Ontario ceased stocking in 1971 and Minnesota in 1974. Some cohos are still occasionally taken in North Shore waters, mostly the result of stocking by the state of Michigan and successful natural reproduction in south shore streams.

The chinook salmon—or king, or tyee—(*Oncorhynchus tshawytscha*) is the largest of the Pacific salmon group and is also a prized sport fish on the Pacific coast. After an initial stocking by the state of Michigan in 1967, the chinook, too, was a great success in Lake Michigan. In the same year, Michigan also planted Lake Superior with young chinooks. Minnesota began plantings of chinook in 1974 on the North Shore and has continued to the present; Ontario has not planted chinook salmon in Lake Superior.

Chinooks average a larger size than coho, up to forty pounds in Lake Michigan. But Lake Superior chinooks also run smaller than Lake Michigan's, with the usual size up to twenty pounds. The lower temperatures and less fertile waters of Lake Superior cannot compare with the high productivity of Lake Michigan in producing large fish. Nevertheless, chinooks now provide good angling off the Minnesota coast, mostly by trolling, in late summer and early fall. The chinook, like the coho, appears to be spawning more successfully in eastern streams of the Ontario shore, working westward.

The chinook's obvious presence in tributary streams in the fall often triggers much excitement among anglers and tourists: a big fish in a very small stream. Although it is true that the stream-running chinook does not actively feed, some skilled anglers have developed techniques for enticing the big fish to strike a small lure—the ensuing battle of big fish in a small stream becoming an affair of great excitement and much splashing on the part of both fish and angler. The record chinook in Minnesota waters, twenty-nine pounds and ten and one-half ounces, was taken from the Cross River in 1985.

The introduction of the third species of Pacific salmon to Lake Superior was an accident and had nothing to do with the sea lamprey or with restoring the sport fishery of the Great Lakes. Nevertheless, the pink, or humpback, salmon (*Oncorhynchus gorbuscha*) became so well established as a naturally reproducing population, and in such great numbers, that it has earned a firm if somewhat dubious reputation in the North Shore sport fishery.

From its initial release at Thunder Bay in 1956, the pink salmon increased, slowly at first, then spectacularly later, to reach population levels that created a nuisance. Like the other Pacific salmons, pink adults die after spawning, and the presence of so many decaying corpses in small streams along the North Shore caused many upturned noses as well as tempers.

The pink, also like its generic cousins in *Oncorhynchus*, does not feed actively once it enters stream water. Furthermore, the pink's body begins to deteriorate rapidly once it reaches the currents of tributary streams. But if taken bright and fresh from Lake Superior, or just as it first enters tributary streams, the pink is delicious eating. North Shore anglers soon developed skills to take the pink—using small spinner lures and tiny yarn flies—from stream mouths and lower stream reaches.

No stocking of pink salmon is done, and all current populations—now spread through all the Great Lakes—have resulted from natural reproduction. For unknown reasons, the pink's population has decreased, especially on the Minnesota coast and in the Thunder Bay area, although in some Ontario streams farther east the run of pinks is still heavy.

Two additional species of exotics enjoying limited success are the Atlantic salmon and brown trout. Both are stocked by Minnesota in selected streams along the northwest coast, basically to add to the diversity of the recreational fishery resource.

The Atlantic salmon (*Salmo salar*) is a large, silvery, hard-fighting sport fish favored by anglers on both sides of the north Atlantic Ocean. The brown trout (*Salmo trutta*), a native of western Europe, is mainly a stream resident and has been introduced widely with great success in suitable streams around the world. The brown has been eminently successful in post-sea lamprey times in Lake Michigan, growing to huge sizes. Both species are now encouraged and stocked on the Minnesota portion of the shore. Neither is stocked or encouraged by Ontario, but "rovers" are sometimes taken by anglers in Canadian waters.

Interestingly, these two species, so different in appearance in their natural habitats, when introduced to inland lakes, either by natural accident or by human hands, change color and appearance to be nearly identical. In Lake Superior, the Atlantic salmon and the European brown trout are virtually indistinguishable.

Beginning in the late nineteenth century, almost every fish species remotely conceivable as a possible introduction was experimentally planted in the Great Lakes, including Lake Superior. Almost all such introductions failed. In the early 1900s, even the Pacific salmons, chinook and coho, were planted, but unsuccesfully—in contrast to their great success later.

In these earlier times, before human enterprise affected native fish populations, it would seem that the natural structure of Lake Superior's fish community, evolved over many millennia, was so well adjusted within itself that strangers from either the Pacific or Atlantic coast found no niches in which they could prosper, and soon died out.

But after the intense commercial fishing in the first half of the twentieth century, and then the scourge of the sea lamprey, the natural structure of the native fish community was broken down. No longer was the community knit by relatively stable populations fully utilizing the lake's resources. A new community, unstable, with fluctuating food resources, was susceptible to invasions of all sorts. Exotic trouts, smelt, Pacific salmons—all found a place, some only temporarily, some wildly accelerating. The success of the chinook, for example, is only one reflection of this instability.

Two schools of thought now conflict over management of Lake Superior's sport fishery.

One school accepts the premise that the Great Lakes fish populations, including Lake Superior's, have already been irreversibly disrupted, and that the obvious policy should be one of trying to make the best of the new circumstances. This effort includes adding new species to replace the old, hoping they will take hold—a sort of shotgun approach. Obviously, some successes, like the steelhead and chinook, have so far been enjoyed.

The other school, with traditional reverence for the natural system, holds to the promise of a return to the native fish community, or close, that had been so well-tuned to Lake Superior conditions in the past. This school admits the demise of the natural system in Huron and Michigan—but not in Superior, where substantial parts of the native fish populations were saved. Here, enough of the original lake trout and whitefish, and their native forage fishes, survived the initial onslaught of the sea lamprey to warrant a salvage operation that might return the stability of the past to Lake Superior's fish community. Only time, hard work, and experimentation—together with the establishment of now unknown and unforeseen ecological balances—can lead to a return to that stability.

And time will tell whether the future thrill of landing a fighting fish in North Shore waters will come from a native trout or an exotic salmon—or some strange hybrid.

# THIRTEEN

# Wings Along the Coast

On a lonely subarctic sand spit, the young Semipalmated Sandpiper picked its way along the edge of lapping waves that rhythmically rose and receded on the sandy shore. The tiny gray-buff bird was almost indistinguishable on the wrack-scattered sand. As a wave receded, the sandpiper trotted toward the water, stabbed once into the wet sand with its short, black bill to capture a minute crustacean, then trotted rapidly back ahead of the next wave, as if it were afraid to get its feet wet.

Then two other sandpipers flew past, low over the waves, and the young bird suddenly leaped up to join them in flight. The three alighted a hundred yards up the beach. Several others were already there, and the coterie continued a purposeful patrolling along the shore, two or three at a time dashing toward a receding wave, feeding on invertebrates trapped in the shifting sand, then dashing back. The invertebrates were common in the shifting sand, washed up with each wave. A sandpiper rarely made a dash after a wave without capturing one or two of the tiny crustaceans.

The time was mid-August, and already the days were noticeably shortening. Each morning, the sun rose more dimly, farther in the south, and each evening twilight darkened a little earlier. Autumn coolness was more and more apparent. The shorter days

*Herring Gull, Coldwell Peninsula*

delivered a clarion call from warm, tropical sea beaches and freshwater lakes, triggering ancient instincts.

On one particularly cold night, the full moon shone clearly, silvering the constantly moving sand and the breaking tops of wavelets. The old instincts became irresistible. And the next morning, when the first brassy tints of sunrise swept the sand spit, the beach was empty.

The Semipalmated Sandpipers had begun their incredible migration southward.

Groups of three, four, and five collected into flocks of hundreds, then thousands. Single, large congregations usually were

217

composed of either young or adults, either males or females. From the Arctic and Subarctic, far north of the Great Lakes, the tiny birds flew in short, daily flights, unnoticed by human occupants of the rare settlements near which they passed. They began an almost invisible journey of 4,000 to 5,000 miles, or more, to reach their wintering ponds and beaches in the warmth of subtropical and tropical climates.

But then one obstacle lay in their flight path. After some 1,500 miles, the sandpipers encountered a vast expanse of open water. To the south, no horizon was visible. With daily temperatures dropping more rapidly and daily photoperiods growing shorter, their migratory urge heightened, but below lay only a rocky, hostile coast: the North Shore of Lake Superior.

For some of the sandpipers, at the Thunder Bay-Nipigon Lakehead, friendly bays and beaches drew them down for welcome, albeit temporary, respite. And farther east, the way was clear, almost directly south, toward the great lake's outlet at Sault Ste. Marie. The tiny birds followed a diverse and complex shoreline of granite headlands interspersed with many familiar sandy shores.

Those sandpipers arriving west of the Lakehead, however, encountered a much different kind of shore—inhospitable lava cliffs and rocky spurs. But the shore angled southwestward and, as down the side of a funnel, the sandpipers began to follow its course.

For 100 miles they followed the rocky shore; only a few small grassy beaches, their usual habitat, lay below. There appeared no end to the hostile coast. But gradually, the dim line of a thin horizon appeared on the faraway edge of water to the south.

The open lake soon came to an abrupt end; the wide, island-dotted estuary of a river extended westward and north upstream to rapids. But the sandpipers were not attracted to the St. Louis River and its sources. For, across the mouth of the wide estuary, almost unbroken from north to south spanning the end of the vast lake, was a familiar spit—sandy strands at water's edge, grass and brush-covered beach lines, and, a few yards upland, a line of thick brush and piny forest for shelter from possible storms. This was Minnesota Point, familiar to the adults, appearing hospitable to the young, a welcome stopover in their long journey. The flock descended to the scene of sand and gentle waves.

Here the Semipalmated Sandpipers found a familiar environment—temporarily. They separated into small groups of two or three, or individually. And they began again to follow their ritual of feeding at the water's edge, alternately chasing a receding wave upon the sand, racing shoreward again ahead of the next. Now food items were slightly different: here they were primarily insects—fine sand cases with caddisfly larvae and pupae, tiny mayflies, larval midges—for this was fresh water, not the brine of the northern seas in which insects could not live.

The sandpipers' stay was ephemeral, for signs of approaching winter continued more noticeably. And soon the friendly sandy beaches of Minnesota Point were empty again. Ahead lay thousands of miles more before the winter warmth of tropical Central and South America was to be encountered. The North Shore, with its autumnal storms and wintry, ice-coated rocky coast, was abandoned.

In eastern North American breeding grounds, the Semipalmated Sandpiper—*Calidris pusilla*—is the most abundant small shorebird. But it is only one of several members of the shorebird family, Scolopacidae, that regularly migrate through Ontario and Minnesota, around the western end of Lake Superior, temporarily occupying Minnesota Point. As a group, sandpipers are small birds, known, with some degree of fondness by birders, as "peeps." Some, like the Semipalmated, are best described as sparrow-sized, a scant six inches long. Coloration is obscure, mottled olive-brown on the back; their scurrying movements are the most noticeable signs of their presence. Two other peeps—the tiny Least Sandpiper (*Calidris minutilla*) and the Sanderling (*Calidris alba*)—also breed in arctic and subarctic Canada and winter in the tropics. The Solitary Sandpiper (*Tringa solitaria*) breeds north of Lake Superior, but inland; it is found not along sandy spits but on the edges of woodland pools and rivers, and it uses old tree nests of other birds such as robins and grackles.

Yet another—the Spotted Sandpiper, *Actitis macularia*—includes the strands of the North Shore in its breeding range, the

219

only sandpiper to do so. Not abundant in flocks, more solitary in small groups of perhaps two or three, Spotted Sandpipers can be found throughout the summer on the few spots of grassy shore interspersed among rocky headlands and cliffs. The species is found inland as well, along streams and marshes. In the summer, it is the only member of the shorebird family commonly observed by birders along the shore. The Spotted Sandpiper can be readily identified by prominent black spots on the breast of the adult in spring and summer (although fall plumage on the underparts is white). This species is also characterized by its "teetering": an up-and-down motion of its tail and hindquarters, as it works its way along sloping rocky shores or inland stream edges. The Spotted Sandpiper is capable of diving and swimming, using its wings to "fly" through water, in both lakes and streams. It is said to "flush" directly from the water's surface into full flight, which must be a startling performance. This species, too, abandons the icy shore during the winter in favor of southern climates.

The American Woodcock (*Scolopax minor*)—choice sporting game of the upland bird hunter—is also a member of Scolopacidae. But the woodcock long ago abandoned sandy shorelines as its habitat in favor of upland woods and swamps. It, too, is funneled around the great lake on its way to the swamps of southern Gulf Coast states.

Members of the Scolopacidae are not the only migrating birds that are forced to make use of the funneling effect of the northwest coast of Lake Superior and the hospitality of Minnesota Point as shelter and feeding grounds on their way south. Many small, inland forest birds, nocturnal and therefore unnoticed, in making annual north-south migrations, are also funneled around the western end of the lake. Minnesota Point—a "migratory trap"—is a favorite of birders in the autumn.

Most noticeable of all is the annual pageant of migrating hawks, eagles, and falcons passing over Hawk Ridge, at Duluth. The ridge itself is formed of resistant diabase sills, once lying at about the level of old Glacial Lake Duluth, now part of Duluth's Hawk Ridge Nature Reserve. The reserve is an area of more than 300 acres, set aside for public use and enjoyment of this annual avian spectacle.

Flying during the day, especially with favorable winds that follow cold fronts, taking advantage of thermal air currents up

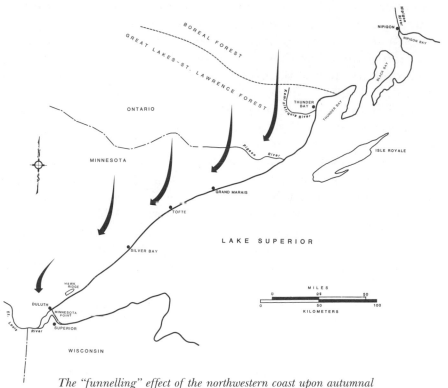

*The "funnelling" effect of the northwestern coast upon autumnal migrations of birds. Many species become concentrated at such locations as Hawk Ridge and Minnesota Point while attempting to pass around the western end of the lake.*

the steep hillsides, many thousands of migrating birds concentrate over Hawk Ridge in their sweep around the end of the lake. Whereas the bird migrations past Hawk Mountain in Pennsylvania, or New Jersey's Cape May, might be better known, Duluth's Hawk Ridge boasts the highest recorded autumn total of hawks in the continental United States—an annual average of 50,000, with a record of almost 75,000 in 1978. One daily total of 33,000 hawks, mostly broad-wings, was logged on a mid-September day in the same year.

Huge flocks, or "kettles," of broad-wings soaring over the high bluffs bring excitement to groups of enthusiastic birders. Beginning in 1951, recorders annually tot up the species and

221

numbers from observation points along the ridge. Some capture and band birds for migrational studies, 2,000 to 3,000 each year.

Whereas Broad-winged Hawks make up the great majority of raptors funneling down Superior's shore and around the lake, many others are also recorded: the Sharp-shinned Hawk, Golden and Bald Eagles, Ospreys, Kestrels. Turkey Vultures, making up one of nature's clean-up squads for removing carrion, are also common at Hawk Ridge in the fall.

The rare Peregrine Falcon, which once nested in its aeries of North Shore cliffs, has been recorded at Hawk Ridge, but its last native presence along Lake Superior was in 1964, a notable victim of DDT poisoning. In the past, the peregrine was rare enough on the North Shore—at most, five or six nests, in the Silver Bay to Grand Marais area—but its sky-high, predatory aerobatics were a wonder of the bird world. Now attempts at reintroducing the peregrine are under way along the Minnesota coast. In 1984 and again in 1985, banded, radio-carrying birds were released by University of Minnesota biologists on a cliff near Tofte, and initial results were highly successful. Soaring in thermal updrafts along the lava cliff, the peregrines seemed to enjoy their return to an ancestral home.

The autumn bird concentration over Hawk Ridge is one of nature's great migrational displays. It remains unequaled on the North Shore. And soon, perhaps, locally nesting Peregrine Falcons will be included again in annual tallies.


Although the inland bird fauna north of Lake Superior is large and diverse—with well more than 200 species—the fauna along the shore itself is sparse in both diversity and abundance. Dependent upon fish and other aquatic life produced in Lake Superior's few shoals and rocky edges, bird life here is a product of the infertile character of the great oligotrophic lake. The fame of Lake Superior's fisheries rests mainly upon seasonal concentrations that give only the illusion of great productivity, and, in the main, the waters of the northern shore are clear and nearly empty of aquatic life. The few bird species regularly inhabiting or breeding on the shore itself, relative to the inland avifauna, reflect this sparse fish abundance.

Nevertheless, the human visitor cannot but be impressed

with the apparent abundance of such birds as the gulls—
"seagulls" to many visitors—primarily the Herring Gull and
Ring-billed Gull. With their constant mewing cries and wheeling
flight over rugged headlands and rocky skerries, the gulls present
an image of a wild seacoast. Feeding on fish offal, the concentration of gulls near commercial fishing operations presents a teem-
ing activity. Waiting on breakwaters and piers for the return of
fishing boats, constantly patrolling developed harbors and mari-
nas, and flocking to tourist waysides, the gulls add their unique
flavor to the sounds and images of the North Shore.

The Herring Gull (*Larus argentatus*) is the most common. It
nests on treeless, remote shorelines and islands, where nests and
eggs are usually predator-free. The Ring-billed Gull (*Larus dela-
warensis*), nesting on granite-based islands in northern and eastern Lake Superior, is smaller and less common. Interestingly, the
ring-bill's population appears to be increasing rapidly at the western end of the lake, on some dredge-spoil habitat in the Duluth-
Superior harbor.

The populations of several gull species suffered declines in
the late 1800s, throughout much of the northeastern part of the
continent, owing to a unique combination of factors. Some species—notably the Ring-billed Gull—were brought close to extirpation by a triple threat of unregulated forces: egging, market
hunting for sale, and hunting and trapping for feathers to serve
the millinery trade. With nests on islands where predators could
not follow, but on the ground with open visibility, the gulls' eggs
were particularly vulnerable to human predators, who could
reach the islands in boats.

But a groundswell of both public and professional concern
for birds arose in the last two decades of the nineteenth century.
The American Ornithologists' Union was founded in 1883, and
the first Audubon Societies formed in New England in 1896.
Soon the rush to join the bird protection movement was a land-
slide. State laws were passed outlawing market hunting and egging, and some against the millinery trade, too. In 1900, the federal Lacey Act prohibited interstate transport of birds killed in
violation of state laws. And in 1918 the Migratory Bird Treaty between Canada and the United States was ratified. Gull populations were soon on the rise.

Today, the Herring Gull and Ring-billed Gull are the two

main gulls in well-established populations on the North Shore. The ring-bill, apparently once having been near extirpation along Lake Superior's shores, continues to increase.

Both gulls depend on fish or fish remains for their staple diet. When declines occur in fluctuating fish populations, so too the numbers of gulls decline, through reduced reproduction. Gulls migrate south in late autumn, although some will remain through winter if open water and food availability permit. The fall gull migration is not simply a directional flight south, however, but appears to be a general dispersal to large, open waterways such as the other Great Lakes, the St. Lawrence River, and the Atlantic coast.

In Duluth-Superior harbor, along newly created shorelines and beaches, now breed some birds not found commonly in this part of the lake, such as the ring-bills and Common Terns (*Sterna hirundo*).

The Double-crested Cormorant presents a striking counterpoint to the gulls. Flying high in V-shaped formations, searching an inland river for food, or waiting patiently in lines on one of Thunder Bay's breakwaters or piers, cormorants are fishers of a totally different image. In strict military precision, these big, black, seemingly ungainly birds stand at attention in line. With upright, correct military posture, they have been called "soldier birds." Diving to great depths, they are able to spend a long time under water in search for prey fish.

The Double-crested Cormorant—*Phalacrocorax auritus*—is in a bird family separate from the gulls, but like the gulls, the cormorant seeks isolation for breeding and nests on rocky islands. Once nearly extirpated owing to pesticide contamination in their fish food base, and subsequent lack of successful reproduction due to eggshell thinning, the Double-crested Cormorant is now making a slow comeback. It, too, migrates in autumn, spending the winter along the warmer, more fertile, fish-producing coasts of the Gulf States.

Adding special flavor to the North Shore are the rare appearances of migrating waterfowl outside their usual range, such as the King Eider (breeding in the far Arctic), and Barrow's

Goldeneye and the Harlequin Duck (both of which characteristically breed on the Pacific coast *and* Labrador).

Among waterfowl taking up regular summer residence on the North Shore, the mergansers are exemplary. Three species, Red-breasted (*Mergus serrator*), Common (*Mergus merganser*), and the diminutive Hooded Merganser (*Lophodytes cucullatus*), are all fish eaters. All breed along the North Shore but migrate south for winter. The Common Merganser, however, will often be found in open water in winter. On the other hand, in summer the Red-breasted Merganser will usually be the most common duck observed along the shore. Of the three, the male Hooded Merganser stands out in appearance, with his high crest of black and large white triangle on the side. Among the three, the hooded's preference is more for inland waters and streams; the speeding flight of a male Hooded Merganser up a North Shore river valley is a spectacular sight.

Ranging inland as well as along the shore, all mergansers appear to delight in the rapids of swift streams. They are all diving ducks, feeding on fish along the shore, although inland they feed mostly on other aquatic animals, such as frogs, crayfish, and insects. Along rocky shores, mergansers are mainly solitary, appearing to enjoy in isolation the turbulence of wave and surf, but family groups are common in protected coves and rocky grottoes. Disdained by the hunter because of their strong, fishy flavor, the mergansers nevertheless contribute an ever-fascinating component to Lake Superior's wild shores and inland streams.

A thunder of wings underfoot, the sudden spray of yellow aspen leaves, an escaping shadow through spruce branches—all in something less than two seconds—is a startling but thrilling event. It is the flush of the Ruffed Grouse.

The experienced upland gunner is galvanized to rapid action, instinctive, skilled. With some luck added in, the hunter's snap shot will add a grouse to the day's bag—and a delicious entree to tomorrow's dinner. The novice, however, will probably simply be stunned by the explosive flush and frustrated, unable to draw a bead on anything more than a falling aspen leaf that marked the bird's flight.

The Ruffed Grouse—*Bonasa umbellus*—is the most widely

distributed and most abundant upland game bird in North America. Its range extends from the Atlantic to the Pacific, from Alaska to southern Appalachian Mountains, providing sport to more hunters than any other species of game bird, in both Canada and the United States. And so likewise does the Ruffed Grouse to those who tramp the swamps and creek valleys of the North Shore.

Nevertheless, the Ruffed Grouse is basically a southern bird. Its principal habitat is deciduous woodland, including bushes, aspens, and fruit trees that provide seeds, buds, and berries. It is a bird best adapted to a disturbed forest in varying successional stages. Like its ecological associates, the white pine and whitetailed deer, the Ruffed Grouse is very much at home in the Great Lakes-St. Lawrence Forest. Thus, along the North Shore, it is most abundant in the mixed woodlands at either end of the lake.

Normally, the Ruffed Grouse on the ground becomes nearly invisible as prey. In its plumage of mottled brown and gray, with its slow and deliberate movements, it blends imperceptibly with a shadowed forest floor. In a tree, where it often roosts at night, the grouse displays an uncanny ability to visibly become part of bark or branch. But in winter, grouse provide a major source of prey for many of the northern predators—the Northern Goshawk, several owls, the lynx, fox, and coyote. At this time, goshawks— probably the Ruffed Grouse's main predators—migrate south from the subarctic. And along Lake Superior the presence of conifers provides more cover for the goshawk than for the grouse. But here also deep snow provides the grouse's escape from extreme cold at night, by diving into snow.

Farther south—for example, in central Minnesota—courtship "drumming" reaches a peak about mid-April, but it is late April on the North Shore, a week later. Unfortunately for the grouse on the North Shore, drumming when snow still covers drumming logs and escape coverts exposes the grouse more visibly to predation.

There appears little reason for concern about natural mortality by the grouse's predators, for these communities have evolved over many millennia, to reach the present status of sustained productivity. In fact, the richness of faunal diversity on the North Shore would be poorer indeed were it not for the Ruffed

Grouse, through predator-prey interactions, serving to convert the energy in seed and bud to animal flesh, and then to relay this nutriment on to its mammalian and avian predators.

Given the grouse's cryptic coloration and habits on the ground, and its startling and swift flight, even heavy hunting pressure does not appear to endanger bird populations. The skill of even the most experienced handler of shotgun cannot equal that of the hunting goshawk. Most grouse cover north of Superior, remote and roadless, is rarely visited by a hunter, and some may not ever be.

Grouse are not all the same around the North Shore. Where the mixed coniferous-deciduous woodlands of the Great Lakes-St. Lawrence Forest change to dense conifers of the Boreal Forest, distinct changes in grouse populations occur, too. For, where white pine and hardwoods give way to black and white spruces, not only does the abundance of grouse change, but the species changes, too—from primarily Ruffed Grouse of southern preferences to the Spruce Grouse of northern conifers.

The Spruce Grouse—*Dendragapus canadensis*—is a close relative of the Ruffed Grouse, and its geographical distribution is also very broad, from coast to coast. But it differs from its ruffed cousin in some most remarkable ways. For example, as a bird well adapted to the spruces of the boreal forest habitat, its winter food is a monotonous diet of spruce and jack pine needles, in contrast with the Ruffed Grouse's seed, bud, and berry diet. Unlike the Ruffed Grouse—in which hen and cock are virtually indistinguishable —the Spruce Grouse shows a strong sexual dimorphism, and anyone seeing the cock Spruce Grouse in full strut in the mossy bower of his spruce swamp will surely be enthralled at the bird's rich display of color—black and crimson and white in striking contrasts. On the other hand, hen Spruce Grouse resemble the Ruffed Grouse in overall appearance, a mottled gray and brown.

Spruce Grouse occupy coniferous woodlands in the vast transcontinental boreal forests of Canada, southern Rockies, and sub-alpine habitats of the North. In contrast to Ruffed Grouse, Spruce Grouse populations are sparse; but then neither do they fluctuate in the well-known "cycles" of Ruffed Grouse. Spruce Grouse may migrate seasonally to some extent, probably an adaptation to extreme cold in mountains or far northern forests, whereas Ruffed Grouse are solid year-round residents.

The climax forest of mature coniferous trees is the principal habitat for the Spruce Grouse, and when this habitat is destroyed by timber harvesting or other human activites, the loss is also felt by the Spruce Grouse. But the following second growth, especially if aspens are included, favors the Ruffed Grouse. And so it was around the upper part of the North Shore. For, although the Lakehead area was part of the Boreal Forest region, human activities here transformed grouse habitats near Lake Superior to favor "Ol' Ruff." In logged clearings and along forest trails, Ruffed Grouse now prosper and are the main object of bird hunters in this region. The Ruffed Grouse was not as abundant in prehistory.

The overall ranges of the Ruffed and Spruce Grouse overlap completely on the North Shore, but only in gross dimension. Spruce Grouse may be found occasionally on the Minnesota portion, where isolated spruce swamps create their habitat. And Ruffed Grouse occur locally in Ontario's boreal forest, especially now near Lake Superior shores. Locally, habitats of the two birds are sharply defined and do not overlap.

Other members of the avian fauna have separated out in similar fashion. For example, the Blue Jay is common in the deciduous and mixed forests; in coniferous forests, the Gray Jay—the "camp robber"—is its ecological counterpart. In like distribution are the Black-capped Chickadee in mixed forest, and the Boreal Chickadee in conifers. And still more pairs of ecological types can be found among North Shore birds, all reflecting the strong relationship between habitat and speciation—like Darwin's finches in the Galapagos.

One other difference between the two grouse is notable, leading to a sobriquet frequently applied to the Spruce Grouse: fool hen. Lacking almost totally in wariness toward its human observer or pursuer, the Spruce Grouse will sit or walk about under the closest of scrutiny and permit the most intimate approach.

Interestingly, such lack of caution is exhibited only toward humans. Against other predators and bird dogs, the Spruce Grouse presents the same thunder of wings and fleeing shadow as its ruffed cousin, in the gloom of its boreal forest.

In springtime, another thunder echoes through Ruffed Grouse habitat: the rolling tattoo of the male drummer in courtship. The Spruce Grouse, in feeble imitation, produces a drumming, too, but it is barely audible and consists of only two or three beats.

It is time to renew the population.

So the Herring Gulls return to nest on protected crags, cliffs, and islands, and Spotted Sandpipers to find their few grassy beaches along the North Shore. Hawks and owls, on their north-ward migration not finding a convenient funnel of lake shore to guide them, arduously make their way back around or across the broad inland sea.

And on a subarctic sand spit, now in early June relieved from the harsh desolation of winter's cold, the days of sunlight and warmth lengthen. And one morning, as an early sun breaks upon the wavelet tops now with a golden, lambent light, a group of three Semipalmated Sandpipers appears. Immediately, they race the receding waves, stab the shifting sand for food, and race back. They must feed rapidly and intensely, to obtain protein for egg production and to gain strength for a later return flight.

Their season will be short.

# FOURTEEN

# *The Antlered Mammals*

Wilderness symbol of northern tundra and boreal forest, the ancient caribou is one of our few survivors from the Ice Age fauna that followed the retreating glaciers. It survives still, on isolated islands and in remote forested areas, along the North Shore.

In the primeval boreal forest the woodland caribou was the principal large mammal, the ancient species that provided food and clothing for the humans of North Shore prehistory. It also provided food for the gray wolf, its main predator, in a long-established mutuality.

Only a scant century ago, the caribou remained the common large game animal of the region, the only antlered mammal in the vast area north of Lake Superior. Moose and white-tailed deer lived to the south—both east and west of the Great Lakes—but by the year 1850 they had not extended their ranges around the Great Lakes.

Civilization advanced into the mixed forest on the western North Shore, in the latter 1800s, and a half-century later, to the boreal forest farther north. The climate changed—for the warmer—during this past century, and moose increased northward, dispersing from both ends of the shore. With the clearings and fires that followed the logger's ax and settler's plow came also

*Young moose, Lake Nipigon area*

the white-tailed deer—and with the deer an insidious disease to which the caribou had no resistance. And as the deer increased, so too did its principal predator, the gray wolf, which found the caribou an even easier prey than the deer. Hunters also brought their guns—perhaps the most devastating factor of all in caribou losses.

And so the caribou retreated and declined. In its wake followed in turn the moose and white-tailed deer. Together, these three wild mammals now constitute the main fauna of Cervidae, the deer family—those that annually drop and renew their antlers—in the Great Lakes region.

The litany of their fluctuating populations and distributional succession is an ecological classic. Principles underlying their interrelationships are nowhere better exemplified than in the drama of these legions of antlered mammals that waxed and waned in succession along the Superior North Shore, a succession that is still dynamic and changing.

The drama began to unfold 18,000 years ago with the start of decline in the Wisconsinan glacial stage. The edge of ice retreated northward, and following came tundra and boreal forest. Ice Age humans dispersed northerly, too, trailing the large game animals that had crossed Beringia. This process of change was long—or seemingly so to us today, looking backward with our retrospective studies in archaeology and paleoecology. The megafauna disappeared to extinction, and by the time Europeans penetrated the North Shore, only 350 years ago, they encountered the Ojibway living in apparent stability with the caribou as their staple big game.

The process of change was about to accelerate dramatically.

The caribou—*Rangifer tarandus*—has a circumpolar distribution; it crossed Beringia into North America from Asia along with many other mammals of ancient lineage. Three major populations—the reindeer of northern Eurasia, the barren-ground caribou of arctic North America's tundra, and the woodland caribou of the boreal forest are grouped variously into several subspecies. Only the last of these three—*Rangifer tarandus caribou*, the woodland caribou—need capture our attention on the North Shore.

The woodland caribou is among the largest of the caribou, with bulls weighing up to 425 pounds. Pelage is gray-brown, lighter on the belly, often ivory on neck and shoulders, with a whitish beard and a prominent light-colored rump patch. Unlike most others in the deer family, caribou of both sexes have antlers, the bull's being larger. One of the bull's palmate tines, modified larger than the others, extends forward over the middle of the brow and gives the appearance of a single, separate structure, known as a "shovel tine." With its covering of thick, buoyant hair, the caribou is an animal with a remarkable swimming facility and an ability to withstand cold. It is well adapted to the lakes and rivers of the boreal forest and the northern winter. With large, crescent-shaped hooves, it can traverse spongy muskeg as well as deep snow.

Most interesting, perhaps, is its diet that includes lichens of mature conifer forests. These include both tree lichens growing on dead limbs of spruces, jack pine, and balsam fir ("old man's beard") and ground lichens ("reindeer moss"). In summer, wood-

232

land caribou browse on herbs and deciduous leaves, but in winter they feed heavily upon the lichens.

Feeding on tree lichens, made more available when winter storms blow down dead trees or the lichens themselves, or pawing away snow with their large hooves to reach ground lichens, woodland caribou are singularly adapted to the use of this food element of the boreal forest.

While thus enjoying this advantage over grazing or twig-browsing animals, woodland caribou also find a disadvantage in this winter diet. For the boreal forest requires a long time to reach the mature stage producing lichens, and lichens themselves require extremely long periods—up to fifty years—for full development.

In presettlement times, wildfires were a natural part of the forest system. Set by lightning, such fires were random in distribution and left a forest mosaic of trees in all stages of succession, from new growth to climax. Indeed, wildlife biologists now believe that fires opened up the canopy in patches of dense woodlands and thereby fostered ground lichen growth. But a consequence is that the woodland caribou was forced to be nomadic, on the move—not like the mass migrations of the barren-ground subspecies of the tundra, but in smaller family groups—in order to constantly find new supplies of food. It requires a large range.

Through the nineteenth century, climate in the Great Lakes area warmed significantly. And it was in the late 1800s and early 1900s that logging of the North Shore region began, to continue for several decades. Wildlife scientists are uncertain which factors—climate, forest disturbance, disease, predation—were primarily responsible for the caribou's decline. Probably a combination of factors, depending on original forest type and other animal species, was most accountable. Shooting appears to be an important factor in their decrease, for caribou are unwary, docile animals around humans.

So a dramatic decline in caribou populations took place on the North Shore, from south to north as climate warmed, from both east and west as loggers, settlers, hunters, and new large mammal species converged around the North Shore.

It was a decline from which the caribou was not to recover.

Except for protected remnant populations, the woodland caribou on the North Shore was essentially extirpated.

Ten thousand years ago, when late glacial ice was receding northward, the moose—*Alces alces*—occupied several large but isolated areas in the Great Lakes region. Like the caribou, the moose has a circumpolar distribution, and it, too, had crossed Beringia from Asia.

As ice retreated, these several refugial populations of moose remained remote from each other. Two of these concern us here. One, generally in the area of Pennsylvania, consisted of a separate subspecies, *Alces alces americana*; another area, west of the ice that still covered the Michigan basin, held the subspecies *Alces alces andersoni*.

South of the Michigan ice at that time was a broad prairie grassland—an ecological barrier just as effective as the ice. But when glacial recession continued northward, so followed the two subspecies of moose, with the Great Lakes between. And along the northern crescent of the North Shore of Lake Superior was another barrier—boreal forest, unsuitable for moose in its original condition, but abundant with woodland caribou.

The moose is much larger than the caribou, prime bulls weighing up to 1,200 pounds. Like the caribou, moose can survive the cold of winter and are good swimmers. About the size of a horse, moose have long legs and a horselike, pendulous nose. Its hair is dark brown overall, almost black, with a prominent dewlap hanging from the throat of both sexes. However, only males have antlers—large, palmate, and spreading up to five or six feet. The moose's long legs make it appear awkward on land, but they aid in wading lakes and ponds and traveling in deep snow.

Its feeding habits—and therefore its habitat—are different from the caribou's. The moose is a browser on leaves and aquatic plants in summer and twigs and buds in winter; it is thus an animal of the mixed deciduous and conifer forest that followed the boreal forest northward. The moose thrives on forest disturbance, such as fire, feeding on second-growth trees and shrubs in early and middle successional stages. Unlike the caribou, the moose is more solitary and rarely travels in groups.

Thus it was, at the end of the nineteenth century, that the stage was set for the second act in ecologic succession.

Theories differ as to why the moose did not colonize the area around the northern coast of Lake Superior until about 1900. After all, it had had plenty of time: since the end of glaciation. Several possible explanations have been put forth. First, climate had warmed during the nineteenth century, slightly but significantly, favoring the southern forest and the northward dispersal of moose. The warmer climate, of course, disadvantaged the caribou. Second, about this time, the loggers with ax and saw began to change the landscape drastically in the mixed forest with tall pines along the northwestern coast. Fires followed cutting and, especially along the shore, so did human settlement. Deciduous trees and shrubs invaded the logged and burned pinelands. Later, especially in the 1920s and 1930s, mature coniferous trees in the Boreal Forest Region were much reduced, and second growth followed.

The arrival of the moose worked to the disadvantage of the caribou, in some areas, through another most interesting relationship with the predator that preyed on both—the gray wolf. Basic to their relationship were the relative swimming abilities of caribou, excellent swimmers, and wolves, which seem to abhor water. Before the moose arrived, caribou found refuge from wolf predation on islands—in Lake Nipigon, for example—during spring calving periods. Wolves are not good swimmers, and they were not inclined to attempt the swim to islands during times of open water. Islands, then, in spring and summer were safe. During winter, when the wolves could travel over ice to islands, caribou disappeared into mainland forests; no prey were left on islands to support wolves through the winter. When the moose arrived around 1900, however, this system changed. For moose, once they had dispersed to an island large enough to provide good habitat year-round, stayed year-round. Now wolves could stay year-round, too, supported by moose in winter as well as in the warm months with open water. Consequently, islands no longer were safe calving grounds for caribou in spring and summer. Caribou, easier prey than moose, were eliminated in much of their original range. Today, only islands without moose—and wolves—remain to support major caribou herds.

Other intricate details of succession no doubt evade us so far.

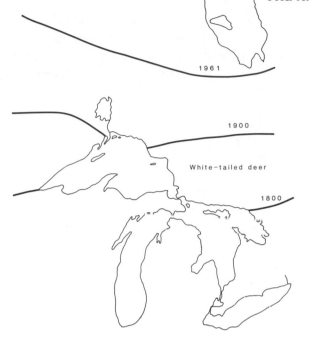

1961

1900

White-tailed deer

1800

*Distribution limits for woodland caribou, moose, and white-tailed deer as human activity moved northward around the Great Lakes in the 1800s. On the maps, lines represent the southern distribution limits for woodland caribou; northern limits for moose (currently farther north); and northern limits for white-tailed deer. (Redrawn from de Vos 1964, with permission of the* American Midland Naturalist.*)*

And it is uncertain whether the caribou disappeared first, allowing space for the moose invasion, or whether the moose came first, outcompeting the caribou. Overall, succession was probably not that simple, anyway. More likely, displacement interactions took place in different local areas, at different times.

In the main event, the caribou nearly disappeared, while the two groups of moose—waiting in the wings, so to speak, at either end of Lake Superior—dispersed east and west respectively along the shore in a great pincer movement, not to close the final gap in the Lakehead area until about 1900.

For the moose, it was a new habitat to be filled, and its range was extended substantially. We are not sure today exactly why and when the two subspecies formed; nor do we know precisely which routes dispersal followed to reach both ends of the North Shore. It seems they invaded the mixed forest in the west earlier than the northern boreal forest. But moose populations in the region north of Superior now apparently represent an intergrade of *americana* and *andersoni*.


The third actor in this drama of succession — the white-tailed deer, *Odocoileus virginianus* — is a native American. It does not occur in the eastern hemisphere. The white-tail appeared in North America perhaps twenty-five million years ago, surviving the Ice Age well south of the glaciers.

Smallest of the three deer, the adult white-tail usually weighs about 250 pounds, although a rare buck in top condition may go as high as 500 pounds, larger than the caribou. Color of the white-tail's coat changes strikingly with the season; winter pelage is drab brownish gray, and the coat in summer is a much brighter reddish brown. Underparts are always white. The underside of the tail is snow white, and when alarmed, the white-tail flashes a warning signal with a highly visible, erect tail. Only the males have antlers; sometimes in a particularly large buck the antlers will create a huge basketlike rack, with many tines. The white-tail wins all honors among the three deer for grace and agility, bounding over fences with apparent ease.

The white-tail is an animal of warmer climes, a browser on deciduous shrubs and trees as well as a grazer on herbaceous plants. In prehistoric times, it prospered among the more agrarian Indian cultures in the east and south and furnished their principal big game, but it was never mentioned by early explorers on the North Shore. The white-tail is the most successful of the deer family, and it thrives in the widest diversity of habitat.

The deer adapted readily to the constant forest disturbance caused by timber harvest and agriculture. Increasing its range northward with new human settlement, the white-tail perhaps reached the area north of Lake Superior about 1900, similar to moose, or at least soon after. When logging and fires brought about the consequent deciduous second growth, white-tail populations increased — exploded, actually, during the 1920s and 1930s — in the mixed coniferous-deciduous areas at both ends of the North Shore.

Using the same foods—second-growth browse—the white-tailed deer and moose differ in their ecological requirements in two important ways. First, the moose with its long, powerful legs can travel and survive in deep snow where the white-tail cannot; and second, the moose can withstand the intense northern cold much better than can the smaller deer.

In severe winters of cold and deep snow, white-tails "yard up" in high concentrations. This behavior is an antipredation migration, occurring to some extent every year. Here in the yard, on packed snow and runways, deer can run and safely escape wolf predation. Along much of the steep topography of the Minnesota coast, the shore provides long, narrow yards.

But starvation occurs when food supplies are cropped too high on tree branches, and when travel outside the yard becomes impossible in deep snow. Consequently, while both the moose and white-tailed deer depend for food on the forest changes caused by logging and fires, the two species sort themselves out by climate—moose to the north, white-tails south.

Recent increases in the number of wolves during a period of severe winter with deep snow—when the wolf had an advantage—have apparently brought about a decline in deer populations. And with further growth and approaching maturity of new forest trees, white-tails will probably never again reach the former high levels of several decades ago. Those were times that today's deer hunters now remember as "the good old days."

Playing a significant role in the succession of these antlered mammals was a tiny, thin, internal parasite—the meningeal worm. It was invisible and unknown until disclosed recently by modern science. The meningeal worm is a parasite with several life-history stages in different animals, including land snails as an intermediate host, as well as any one of the three species of deer. As a larva in the land snail stage, it is picked up by feeding animals—moose browsing along the ground or caribou feeding on reindeer moss—subsequently entering the mammal's digestive tract. Migrating from the new host's stomach, the larva infects the bloodstream, the spinal column, and brain.

In moose, the meningeal worm causes "moose disease": blindness, loss of muscular coordination, and paralysis, eventually

239

death. The disease was observed in Minnesota moose as early as 1912, but the connection to the parasite was not made until 1963, when a Toronto biologist established the relationship.

Significantly, the meningeal worm is carried by white-tailed deer with little physical effect. Consequently, otherwise healthy deer entering the moose's range will transmit through its feces the worm to snails on the ground and then to moose.

So as white-tail dispersal continued northward around the North Shore, following an increase in second-growth food, moose populations declined, victims of "moose disease." Deer and moose coexist only where the deer population is low. Attempts to reintroduce moose where white-tails occur commonly have been unsuccessful.

Recently we have recognized that woodland caribou are even more susceptible to meningeal worm than the moose are, a fact that probably played an important role in caribou decline.

And so it is that today—like the white pine and black spruce, the Ruffed Grouse and Spruce Grouse, the jays and the chickadees—our three deer, woodland caribou, moose, and white-tail, have sorted themselves out along the North Shore of Lake Superior. Moose occupy the cold boreal forest of deep snow that, while much logged and burned, still surrounds the Lakehead region. Here, deer occur only in lake-edge forests, where Lake Superior modifies local climate. But white-tails predominate in the mixed second growth at either end of the shore, west of the Nipigon River and at the extreme southeast end near Sault Ste. Marie.

And the woodland caribou, our wilderness symbol of an ancient prehistoric fauna—where is it now? Assiduously protected, caribou prosper only in isolated areas and remnant populations—primarily on offshore islands where logging and major fires did not occur, where long-lived boreal forest trees reach their climax stage, and where long-lived lichens still grow.

Islands also act as refuges from the wolf, competitive moose, and the disease-carrying white-tailed deer. These include mainly St. Ignace Island near Thunder Bay, the Lake Nipigon islands, and the Slate Islands off Terrace Bay, where caribou still thrive.

Nine miles off the coast, the Slate Islands have been the subject of long-range scientific study in island ecology. Numbering about thirty to forty animals in 1949, caribou populations appear much higher today, with the maturing of coniferous forests since the last major fires in 1930. Neither the moose, white-tailed deer, nor wolf occur in the Slate Islands. Perhaps 400 caribou now range this archipelago, secure against predators and disease.

A few scattered groups of caribou also occur in the remoteness of the Pukaskwa peninsula and inland in roadless wilderness to the north. Individuals are occasionally observed even in northern Minnesota, having wandered south from Ontario. Attempts to reintroduce caribou are planned near Minnesota's Little Saganaga Lake in the Boundary Waters Canoe Area, a small section of virgin forest where both deer and wolf numbers are low.

Other mammals also play their part, much the same as in prehistoric times, interwoven in the ecology of the three antlered mammals.

The gray wolf has been the principal predator upon all three species of deer, although wolf numbers have been much reduced by the white man's gun and trap. Now protected or managed as part of the natural forest scene in most of its range, the wolf enacts its predatory role in a dynamic balance with its larger mammalian prey, a fascinating subject for scholars of wildlife ecology. We will hear more about caribou, moose, and wolves—and their island ecology—in chapters on Lake Nipigon and Isle Royale.

The beaver continues to build its dams and ponds among the intricate waterways of the interior. This small furbearer provides food for the wolf in summer, and in turn creates ponds where moose feed heavily on aquatic plants.

The aquatic vegetation, in turn, provides an essential source of the nutrient sodium, leaching from uplands to be taken up by aquatic plants at concentrations up to 500 times higher than in upland woody plants. The interrelationship between moose and beaver-created ponds is an important element in the ecology of the North Shore woodlands.

In new beaver ponds, old boreal forest trees are killed and eventually fall, and thus arboreal lichens in high treetops become available to caribou.

The tale of exploitation of beaver in the fur trade, the most important mammal in North Shore commercial traffic of earlier centuries, is one told earlier in this book. Beaver harvest is tightly controlled now, but trapping continues to provide an important source of pelts and a base for local human subsistence in some parts of the shore.

The drama of recent forest and mammal succession along the North Shore closes for now. It covered a period of time so short compared with earlier millennia, during which much slower postglacial development took place. The mosaic of mammalian predator-prey interaction, food and dietary requirements, and disease involvement is intricate and diverse, and we cannot predict a future outcome.

Today in spring white-tails leave their yards to reach browse unavailable in deep snow; caribou still swim to islands to calve, where the wolf cannot follow; moose give birth at this time, too, protecting calves against the wolf with powerful legs and bodies.

The succession of antlered mammals was a dynamic sequence of natural phenomena, which the white man accelerated, and in which the geography and changing climate of the North Shore played important supporting roles.

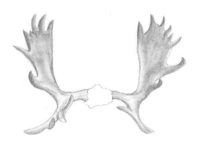

# FIFTEEN

# *The Nipigon Basin*

From a swift reach of stream known as Rabbit Rapids, a short distance down from the head of the Nipigon River, J. W. Cook of Fort William in 1915 landed a brook, or speckled, trout recorded at fourteen and one-half pounds. True, even larger specimens have been reported—found dead, illegally taken, or caught in the commercial fishery. But Cook's fish remains as the world's angling record for *Salvelinus fontinalis*.

The waters of Nipigon—lake and river—became legendary. At the time lacking easy access, remote in a northern boreal wilderness, the brook trout of Nipigon took on a mystique that later river alterations for hydropower and modern highways have still not dispelled.

Nipigon's famed fish fauna included another member that was to receive worldwide acclaim. Antithesis of the big, beautiful brook trout, tiny and ugly, this was the sculpin or muddler—or *cockatush*, as the native Ojibway called it—a bottom-living small fish that inhabits many trout streams. The natives used the ugly sculpin as bait to catch the beautiful brook trout.

In 1937, Don Gapen, an avid Nipigon River angler, noted the effectiveness of live sculpins as bait. He designed a fly of feathers and deer hair, an imitation of the sculpin, and tried it out at Virgin Falls, at the head of the river. Immediately, brook

243

*Diabase cliffs, Nipigon Basin*

trout up to eight pounds in weight succumbed to this early design of the "muddler minnow."

Actually, the sculpin is not a minnow in the technical sense, belonging instead to a separate fish family—the Cottidae. But no matter; the "muddler minnow" of Nipigon River origin is today as famous as the legendary stream that spawned it.

The insular character of large Lake Nipigon and its outlet connection to Lake Superior, the Nipigon River, accounts for the uniqueness of its natural resource history. The region's geological history, though kindred to the Superior Basin, was yet independent, leaving the surface of modern Lake Nipigon 250 feet above Lake Superior. Its early peoples were the Cree, who favored the shores of inland lakes rather than the Lake Superior shores of the Ojibway. Special logging operations were devised to utilize the large lake and its log-driving tributary streams, as well as its river outlet; Nipigon's islands figured in the lasting survival of caribou;

Lake Nipigon's fish fauna, isolated from Lake Superior far below by the river's cataracts and waterfalls for millennia, remains even today distinct in many respects—including record brook trout.

A hundred million years before the onset of the Midcontinent Rift System, a much smaller rift formed in the vicinity of Thunder Bay and Nipigon. By a geological time scale, its activity was short-lived, however, and it did not develop very far. Later it connected to the northern crescent of the larger rift system that was to become the North Shore of Lake Superior, leaving a hundred-mile-long depression, north to south. It probably became filled with freshwater lakes, and over millions of years, silts and sand were washed in to accumulate under water, forming sedimentary rock. The rift today—known as the Nipigon Basin—holds for us a rich treasure of scenic splendor, archeological remnants of human prehistory, and unusual fish and wildlife resources.

A series of rock strata tells the story of the great depression's geological history, all of which at various places are now exposed to view owing to more than a billion years of erosion. At the bottom are the granite rocks of volcanic mountains, formed during the violent time of rifting. Over the granite are thick layers of sandstone and other sedimentary rock known collectively as the Sibley Group, deposited during the Precambrian Era; the name was derived from the sandstone of Sibley Peninsula and Alexander Sibley, one-time head of the Silver Islet silver mining company. Later, magma thrust up from the earth's interior and was inserted as sheets of diabase more than 1,000 feet thick between the Sibley sedimentary layers. And finally, many millions of years later, glaciation left its deposits of gravel and other remnants of glacial drift.

Erosion of the sedimentary "roof" from the diabase has left by far the most obvious and spectacular of the geological formations—the huge, square-shaped mesas and buttes with vertical cliffs still towering as much as 600 feet above surrounding terrain. These features, forming a "badland" topography, are common also along Superior's North Shore in the Thunder Bay area: Mount McKay, Pie Island, Thunder Cape, as well as forming islands in Lake Nipigon and Nipigon Bay. Along Lake Superior's shore south of Thunder Bay, these formations are known as the "Nor'Westers Range," a memorial to the intrepid adventurers of the fur trade.

During late stages of glaciation, the Nipigon Basin existed as a large northern embayment of Glacial Lake Algonquin and was much larger than today's Lake Nipigon. To the north, large proglacial lakes formed against the melting ice in the James Bay area, later to become the saltwater Glacial Tyrrell Sea. Minor bays of Lake Algonquin connected with these proglacial lakes, but the connections were short-lived, and no link between the salt water of James Bay (or its marine life) and Lake Nipigon was ever accomplished.

As ice retreated farther north and water levels lowered, Nipigon's waters became separated at a higher elevation from primordial Lake Superior. A glacial lake formed in the Nipigon Basin, a stage in the draining of huge Glacial Lake Agassiz in the west. Agassiz's waters poured through the Nipigon Basin to the Superior Basin, for 1,000 years, from 9,500 to 8,500 years ago. The glacial meltwater levels in the Nipigon Basin were about 175 to 200 feet higher than present Lake Nipigon. The higher levels formed glacial beaches and terraces, similar to Lake Superior's, still extant above present Lake Nipigon's surface; notable is the distinct terrace on Windigo Bay.

During these last stages of glacial melting, Glacial Lake Agassiz spilled immense volumes of water through the Nipigon Basin and on to the Superior Basin. Major channels were eroded between Nipigon and Superior, later forming the valley of the Black Sturgeon River (now emptying into the head of Black Bay), the Pijitawabik Bay-Orient Bay channel (now carrying Highway 11), and the valley of the Nipigon River itself. Spectacular Ouimet Canyon, south and west of Lake Nipigon, was apparently also carved by glacial waters flowing from Nipigon to Superior, now incorporated in Ouimet Canyon Provincial Park.

By about 7,500 years ago, all of Ontario was free of ice. Meltwaters no longer flowed through the region, and water levels in the Nipigon Basin had dropped to about the present level. It has been Lake Nipigon ever since.

The beginning of the Nipigon River used to be a wild torrent known as Virgin Falls, thirty-five feet high. Now these falls have been obliterated by higher river levels due to hydrodams downstream; the transition from lake to river today is a smooth and quiet glide.

From a now-abandoned Virgin Falls dam, Nipigon's tamed waters flow southward, most scenic in its upper reaches where spruce and cedar-lined shores of islands appear almost everywhere. The busy highway lies to the east unseen and unheard, although a smaller byway parallels the river on the west.

Three more dams downstream impede the river traveler: Pine Portage dam, Cameron Falls at the downstream end of Jessie Lake, and Alexander Falls, all three generating electrical power. Before the dams were built, this stretch of river included Long Rapids, a turbulent reach of two miles that dropped some 130 feet in elevation, necessitating an arduous portage.

Obviously, the Nipigon River is today a conquered version of its once wild, falls-and-rapids character. There is little that could be called rapids now, but much good trout water still exists. The upper reaches—particularly between Pine Portage dam and Jessie Lake—are perhaps the most scenic, fishable, and interesting geologically. Diabase cliffs and their underlying sedimentary rock occur exposed in this stretch, and farther downstream glacial lake clay forms many stream banks, evidence of previous higher water levels in the Superior Basin.

Farther yet downstream, the Nipigon joins Helen Lake, which lies in the ancient glacial valley, extending mainly northward from the entrance of the river. But Nipigon's waters continue south through Helen Lake, and presently enter Lake Superior in Nipigon Bay.

The Nipigon River is the largest river tributary to Lake Superior, and its drainage area is also the largest within the greater Lake Superior drainage. These two conditions, along with its huge source (Lake Nipigon) and its northernmost location, once gave rise to the consideration of the Nipigon as the ultimate headwater of the St. Lawrence-Great Lakes system, as well as to the designation of the Thunder Bay-Nipigon area as the "Lakehead" region. It competes with the St. Louis River, the westernmost major river, for the honorable distinction of headwaters of the Great Lakes. In total, the Nipigon River flows forty

247

miles from Lake Nipigon to Lake Superior, dropping 250 feet in elevation.

Nipigon is Ontario's largest inland lake, measuring about sixty-five miles long (north to south) and forty miles wide, totaling 1,860 square miles; of that total, nearly 1,700 square miles are water and 176 square miles are islands. The lake outline is complex, and so are shorelines, especially when islands are included, for there are approximately 1,000 of them. Fifty-five islands each measure over 0.1 square mile (sixty-four acres) in size, and five are nearly 5.5 square miles (3,500 acres) each; thirty-five interior lakes lie on four of the islands. The highest island, Inner Barn, towers up to 550 feet in elevation above the water surface of the main lake; other bluffs around the lake shore rise up to 600 feet. The Nipigon drainage basin encompasses some 6,000 square miles; the lake lies at an elevation of 852 feet above sea level, or 250 above Lake Superior. Maximum depth recorded so far is about 450 feet. The total length of shoreline, even excluding small bays, is an impressive 580 miles.

The water is cold and clear, even through summer—ideally suited for the coldwater fish species that prosper there.

Explorer Sieur du Lhut, after his visit in 1683, probably borrowing from the Ojibway, called it Lake Alemipigon (though there are many other versions recorded in history). Du Lhut's name for this lake can be interpreted loosely as "water you cross to get somewhere else."

It would seem a good choice. Standing on the shore and staring at a landless horizon or faraway island, an Indian or early traveler would know that he had a mighty *lot* of water to cross to get somewhere else.

The archaeological record indicates that humans have occupied the shores and river mouths of Lake Nipigon for some 4,000 years. The earliest known peoples were thus those of the late Archaic period and the Shield Culture, when copper tools first appeared. Earlier human populations, the Palaeo-Indians who had followed the retreating glaciers into the Great Lakes region, apparently did not occupy the Lake Nipigon drain-

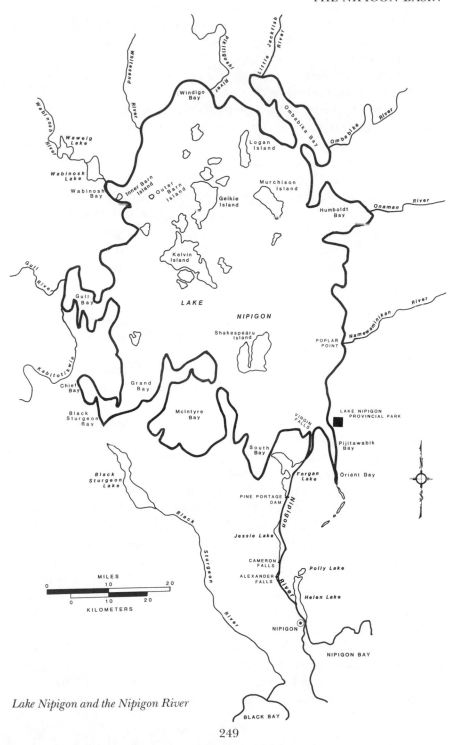

*Lake Nipigon and the Nipigon River*

249

age basin, preferring instead the old shores of larger lakes like Superior and associated glacial lakes, with their flat beaches and elevated terraces.

Human populations were fairly stable and continuous around Nipigon, enduring through the Woodland periods, 3,000 years ago to the time of European contact. In the last decades of these early cultures, the white man's trade goods began to replace native tools and utensils.

The prehistoric Algonkians in the Nipigon Basin, first Cree and later Ojibway, were highly mobile. Loosely organized into multifamily bands, they ranged over large territories. Low food resources necessitated such movements; but these wanderings and inter-band contacts brought on few social pressures, since human population densities were low, and kinship ties existed over many hundreds of miles. The Boreal Forest, homogeneous over vast areas but fluctuating with forest fires, provided much opportunity for hunting, mainly for the woodland caribou. Along the lake's shores and islands these early Americans fished for whitefish, sturgeon, lake trout, and Lake Nipigon's large brook trout.

The Cree were the earliest recorded inhabitants on Nipigon's shores; this northern Algonkian tribe occupied part of the vast Boreal Forest between Hudson Bay and Lake Winnipeg to the west. Even the Dakota, inhabitants of the woodlands around western Lake Superior, apparently occupied some of Nipigon's western shores. After European contact in the east, the Ojibway arrived from their ancestral lands around the St. Marys River to consolidate a position in the Nipigon Basin.

Through the centuries that followed, legends of Lake Nipigon's Ojibway lived on. Told and retold around the fires of winter, one tale is still as chilling as the bitter blizzard that rattles the glass windows of modern cabins, just as it shrieked outside prehistoric wigwams. It is the story of a giant of the northern snows, a monstrous cannibal of the winter woods, with a name that today still strikes terror in the spirit of even the stoutest-hearted: *Windigo*.

The story begins in a common, worldly scene. A single Ojibway, Windigo, alone on the northern shores of Lake Nipigon in the middle of a cruel winter, is starving and weak, for the black spruce forest is empty of game. In desperation, he prays to his spirit Manitou. And in a dream, his prayer is answered: he awakes with his body infused with supernatural powers. With giant strides, he travels south beneath a brittle moon for 100 miles, to Lake Superior's shore. Finding a village, and using his supernatural power with a wild scream, he turns the inhabitants into beavers and devours them all.

Not quite all, however. One lone hunter—Big Goose, a strong and revered woodsman—had been away, searching for game for the village. Upon his return he sees the destruction wrought upon his people and finds Windigo's footprints in the deep snow—giant footprints because Windigo has now grown to an enormous size. Desiring redress, Big Goose prays to *his* Manitou. He, too, grows to a monstrous size—the giant Missahba.

Following huge footprints to Hudson Bay, Missahba engages Windigo in a colossal mountain-hurling conflict. It lasts for two weeks, and the entire land between Hudson Bay and Nipigon shakes with the battle. In the end, the noble Missahba is victorious, and Windigo is destroyed at last.

The story ends happily: Missahba returns as Big Goose, the consumed beavers return as Ojibways again in their village, and only the legend remains of Windigo.

But is the tale all that survives of the Ojibway legacy of legend? Some say that the blood-curdling yell of the snow monster can still be heard in a winter storm on the northern shores of Lake Nipigon. Is it *really* only the blizzard's shriek in the black spruce, along *Windigo Bay*?

The date and names of the very first European explorers and traders up into the Nipigon country remain unknown. But the river route and lake were probably well known by about 1650, undoubtedly by adventurers from New France. The earliest known journey up the Nipigon River by a white man was by Jesuit priest Claude Jean Allouez, on his classic navigation of Lake Superior in 1667. In their village on Lake Nipigon, Allouez reestablished contact with Nipissing Indians from the east, who had been pushed westward by the warring Iroquois.

The first attempt to make economic use of the Nipigon route was by Sieur de la Tourette (brother of Sieur du Lhut), who in 1678 built a trading post on the Nipigon River near the lake's outlet. And five years later, in 1683, the two brothers constructed the first post on the lake itself, Fort à la Maune, in the northeast area near the mouth of the Ombabika River. In the following century, French posts proliferated on the big lake, the route north to James Bay was well developed, and the fur trade flourished.

Hudson's Bay Company remained at its establishments on Hudson Bay and James Bay, in contrast to the more aggressive French. French posts in the Nipigon country served also to intercept Indian brigades taking pelts northward to James Bay—the rivalry between French and English traders at this time was intense and bitter.

But the French and Indian War ended in 1763 with conquest by the English, and the French essentially disappeared from the Nipigon area. The British North West Company, formed in the late eighteenth century, also built posts on Lake Nipigon, and now it was the turn of the Nor'Westers to engage in intense rivalry with the Hudson's Bay Company. By 1821, the two companies amalgamated under the Hudson's Bay Company banner. At this time, eleven fur trading posts were active on Lake Nipigon.

The earliest passage by traders to Lake Nipigon from Lake Superior was directly up the Nipigon River, a difficult, treacherous travail upstream against foaming rapids and waterfalls. An easier route was later established up the eastern valley carved by glacial meltwaters—Helen Lake, Orient Bay, Pijitawabik Bay. North across the big lake was much "water you cross to get somewhere else," and that track was largely a matter of island hopping, making use of prehistoric Indian sites on the islands.

Two major staging areas, with trading posts and habitations, were established on the north side. One was on Ombabika Bay, at the north end of which the Little Jackfish River enters; the route to James Bay was up the Little Jackfish, portage over the Lake Superior-Hudson Bay watershed to the Ogoki River system, tributary to the Albany River, and thence down the Albany to James Bay.

Since it was at its James Bay posts that the Hudson's Bay Company awaited the Indian brigades bearing furs from the Nipigon country, this route from Lake Superior was the scene of greatest conflict between the company and the French, and later between Hudson's Bay and the North West Company, in their competition for the Indians' wilderness harvests.

Hudson's Bay Company (with its incorporated North West Company) built more posts after 1821, for a total of fourteen that continued to 1937. The major fur trade era had long ceased by then, for the beaver had already been trapped close to extirpation. The Canadian Pacific Railway reached the Nipigon region in 1885; it brought contact and rapid transportation with the rest of Canada, and by the early part of the 1900s permanent white settlement had begun.

The Nipigon River is no longer the stream of rapids and falls that existed when Cook caught his huge brook trout from Rabbit Rapids, and when Virgin Falls at the lake's outlet was a mad torrent. The hydroelectric potential of the brawling river was early recognized.

The first generating facility was constructed at Cameron Falls in 1920; this dam did not much affect the rapids character of the upper river, nor did it raise the lake level. But in 1925, the Virgin Falls outlet was harnessed with a dam that served to control the water level, raising the lake level by 1.33 feet; this control helped maintain a dependable water supply in the Nipigon River to the downstream generating station at Cameron Falls. A second dam and generating station were completed in 1930, slightly downstream from Cameron Falls—Alexander Falls—adding to the hydropower capacity of the Nipigon River.

The construction that has contributed the most toward higher water supply is the Ogoki Diversion. The Ogoki River, located north of Lake Nipigon, is in the Hudson Bay drainage and flows to James Bay, via the Albany. But a dam constructed in 1943 on the Ogoki raises water levels by forty feet and diverts water back southward to connect with the Little Jackfish River, flowing into Ombabika Bay and Lake Nipigon. This diversion has the capacity to increase the annual dependable flow in the Nipigon River by 50 percent—from 8,000 to 12,000 cubic feet per second. The added water in Lake Nipigon also contributes to the raising of water levels by another 1.84 feet. But the Lake Nipigon level is regulated and cannot exceed 855 feet above sea level; when that maximum is reached, say, by above-normal rainfall, the diversion is shut down and the full Ogoki flows again to James Bay. The Ogoki Diversion—from Hudson Bay to Lake Superior—adds

253

about 4.5 percent to the outflow in the St. Marys River. Together with the Longlac Diversion, a water flow of 5,000 cubic feet per second is thus diverted from Hudson Bay.

The development that has done the most to tame the turbulent nature of the upper river is Pine Portage dam and generating station. Constructed in 1950, this dam flooded out former Hanna Lake and formed much larger Forgan Lake, as well as reducing the river's upper rapids to mostly flat water. It flooded out Virgin Falls dam, now open to passage. It also raised Lake Nipigon's waters by another 0.20 feet.

Thus the power of the ancient Nipigon is harnessed and utilized for modern domestic and industrial needs. In total, the three stations produce about 275 megawatts, one-third of the regional power used in the Kenora-Nipigon districts. Nipigon's power contributes to, but does not totally serve, the electrical needs of Thunder Bay and Nipigon. But there are those who decry shore erosion and flooding of archaeological sites.

There are also those who remember, sadly, the Rabbit Rapids and big speckled trout of J. W. Cook's day on the famed river.

For a half-century in the middle 1900s, the logger's ax and logging drives of pulpwood worked to change the face of the basin.

The spruces were the principal species of economic importance, main components of boreal forest, prime wood for paper pulp. Trees suitable for sawn lumber, such as white pine, were rare.

Starting in 1923, logging operations on the mainland brought logs directly to landings on Lake Nipigon: Chief Bay, Humboldt Bay, McIntyre Bay, Poplar Point. Other inland operations used rivers tributary to the lake—the Ombabika, Namewaminikan, and Onaman—down which river drives carried logs to the lake. From these landings and river mouths, logs were boomed and rafted across the lake to the outlet at Virgin Falls. And here the rafts were broken up for the drive down the Nipigon River.

The major hydroelectric dams having already been constructed, the logs were floated around by specially constructed flumes. In Helen Lake, logs were again assembled in boomed rafts for towing to Lake Superior and Thunder Bay paper mills.

Changing technology in midcentury, plus road and rail accessibility, brought an eventual end to the use of the river to transport logs to mills. By 1973, train and truck hauls had replaced the river drives in the Nipigon Basin.

Logging was not carried out on the Lake Nipigon islands to any significant degree. High-quality timber did not occur in economically valuable densities, and problems of transportation from the islands were prohibitive. Only on Shakespeare and Kelvin islands were trees felled, and these operations were minimal.

Disturbance by fire has not occurred on the islands nearly as much as on the mainland, and on some islands continuation of boreal forest species has been partly responsible for the limited success of the caribou, which depends upon climax boreal forest. The islands of Lake Nipigon uniquely reflect some basic ecological principles. Large islands in a large lake, many are remote from the mainland. Some are sufficiently isolated that they hold wildlife species—and certain interspecies associations—unlike those on the mainland.

Lake Nipigon's island ecology constitutes the outstanding example of caribou-moose-wolf interactions described in the previous chapter. Several of the Nipigon islands were ideal in size and location off the mainland to serve as safe calving grounds for caribou in the spring: the caribou could easily swim the intervening distance; wolves could not. And caribou dispersed to their boreal forest habitat on the mainland in winter. But when the moose arrived in the Nipigon area about 1900, this safe relationship ceased, for moose then occupied some Nipigon islands year-round, supporting wolves year-round, too. Some caribou inhabit the mainland around the lake in summer, but only where islands are nearby—to use as a refuge when endangered by their predators.

Moose occupied only the larger islands having a sufficient food resource, and today these hold moose and few caribou. But the small islands, without moose, still serve caribou as a wolf-free environment for its spring and summer use, to which, as in the past, wolves do not swim. Small islands in the northern sector of the lake provide the principal summer habitat for caribou. Hunting for caribou, formerly legal, has not been permitted since 1929.

Hunting for moose, once permitted and considered excellent on the larger islands, is no longer permitted on Nipigon's islands. Moose are common on the large islands, whereas the black bear and white-tailed deer, present on the mainland, occur only rarely on the islands.

Estimates of caribou numbers on the smaller Lake Nipigon islands now range from 100 to 200 animals, in total constituting one of the major remnants of the woodland caribou. Their survival is the result of a unique predator-prey relationship in an insular habitat.

The bird fauna around Lake Nipigon reflects the same species associations found along the Superior shore, and for the same major reason: fishing. Herring Gulls, Double-crested Cormorants, Common Loons, Common and Red-breasted Mergansers—all are fishers along the shores of the lake and islands, breeding on rocky shores. It is possible that the numbers of Herring Gull increased on Lake Nipigon after commercial fishing began, for, as along Lake Superior's shores, the gulls follow the fishing boats and feed voraciously on fish offal. Among the shorebirds are also the Spotted Sandpiper and Killdeer, tripping about on the black sand beaches in search of small invertebrates washed up by Lake Nipigon's lapping waves.

Sport anglers still seek out the legendary brook trout of Nipigon's lake and river. Most brook trout fishing is in the lake, made possible over the long distances involved by commercial outfitters, but brooks occur throughout the length of the river as well. Isolated from Lake Superior's strains for a long time, Lake Nipigon's brook trout appears to vary genetically from its older ancestors in Superior. This fish once carried a separate scientific designation—variety *Nipigonensis*—as well as a different common name, silver trout. Lake Nipigon is one of the few bod-

ies of water in which the brook trout occurred naturally above waterfall barriers in prehistoric times.

Natural reproduction is successful in both the river and the
lake's tributaries and spring-fed shoals, but natural progeny are
now supplemented by stocking of hatchery-reared brook trout,
too. Hatchery fish derive from natural fish stocks in the lake, in
order to maintain the genetic purity of the lake population.

A particularly interesting feature of Lake Nipigon's brook
trout is their unusually high growth rate, with fish reaching up to
three pounds at three years of age. More normal growth of natural stream populations of brook trout would probably be a scant
one pound in three years.

But Lake Nipigon is a productive lake, and the unique stock
appears to have a higher genetic capabilty for rapid growth. Unlike the very soft, infertile waters that characterize most inland
lakes on the Canadian Shield, Nipigon's waters are much more
fertile—a factor no doubt related to the lake's unique glacial history. Annual catch of brook trout may be something over 10,000
pounds.

The principal food of the brook trout in Lake Nipigon consists of other fish, mainly small sticklebacks and sculpins which
abound in the lake and river.

Brook trout are not the principal sport species, however, in
terms of numbers caught. In the lake, the northern pike and walleye, or yellow pickerel, are taken in the sport catch in much
greater numbers than the brook trout. Lake trout, less common,
also are favored targets of the angler.

In the Nipigon River, brook trout again dominate the
angler's interest for most of the stream length, although other
sport species are there, too. In Jessie Lake, downstream but above
Cameron and Alexander dams, the lake trout produces excellent
populations and angling. Downstream from Alexander Falls
dam—obstacle to river-running fish—the relative newcomer, the
migratory rainbow trout, or steelhead, attracts many anglers and
provides exciting fishing, even right under the highway bridge at
Nipigon. The steelhead, of course, cannot ascend the falls at the
Alexander generating station and does not occur above it. Likewise—and fortunately—the sea lamprey has been unable to circumvent the falls and dams from Lake Superior, sparing Lake

Nipigon's famed fisheries the destruction it wrought in Lake Superior.

A commercial fishing industry was established on the lake in 1917, in response to increased food demands during the First World War and completion of railroad access. Fishing boomed in the 1920s, with harvests of over two million pounds per year, and many millions of pounds of food fish were taken.

At the top of the commercial fish list was the lake whitefish, with lake trout a far second. Lake sturgeon were important, too, in the early commercial catch, and so were northern pike and walleye.

But with heavy fishing, the commercial catch declined, drastically, so that currently the harvest is less than one-fourth of previous highs. Other species enter into the commercial catch: some, like suckers, burbot, and lake herring, do not enter the sport fishery. Brook trout are not permitted in the commercial harvest, and those caught in nets must be released.

Nevertheless, the inevitable conflict between commercial and sport fishing has arisen around species that are important in both fisheries, such as lake trout and walleye. Commercial licenses are limited, and a yearly quota is set, although the average annual catch is less than the quota.

For the most part, the fish fauna is the same today as it was when first fished for in 1917. Two species of suckers, lake whitefish, herring, northern pike, and walleye still constitute the major abundant species. The lake sturgeon, a particularly long-lived fish that in Lake Nipigon required more than twenty years to reach sexual maturity, crashed in response to heavy commercial fishing in the 1920s. Several exotics occur today that were not present a half-century ago: a brown trout population is now self-perpetuating in a northern tributary stream; smallmouth bass were stocked in 1923 and apparently have reproduced successfully at a low level. But neither of these two introduced species has made an important impact.

Another exotic, however, has recently arrived in Lake Nipigon and is expected to create a major disruption in the lake's relatively stable community—the rainbow smelt. Smelt have been present in Lake Superior since the 1930s, but they were not observed in Nipigon until 1976; the exact source of this introduction is unknown, but the smelt was probably transferred in bait buckets. Smelt are now common throughout Lake Nipigon and form an important food component in the diet of the major fish-eating sport species: brook and lake trout, walleye, northern pike. Greatest concern is that the smelt will present severe competition to the whitefish, the principal commercial species.

On the southeast corner of Lake Nipigon, along the shores of Pijitawabik Bay, is Lake Nipigon Provincial Park. It was formerly named Blacksand Provincial Park, after a beach of dark-colored sand derived from surrounding diabase highlands. It is an area of particular interest and beauty.

While providing the usual amenities, Lake Nipigon Park also includes access to both the expansive waters of the lake and trails into boreal forest stands of black spruce. Sweeping views enthrall hikers from trails high upon diabase cliffs; sunsets across Pijitawabik Bay can be spectacular.

Lake Nipigon Provincial Park is a quiet park, and the campers' pace is unhurried. Sometimes the loudest sound is the lapping of swells on the rocks of Pijitawabik's shores.

Lake Nipigon is sometimes referred to as the primary, or most northerly, of the Great Lakes. Not generally included as one of the Great Lakes, still it is certainly the largest and northernmost of inland lakes tributary to the Great Lakes. Almost any map of North America shows Lake Nipigon as a major feature of the continent. Important in the prehistory of earlier human cultures, significant to the fur trade that in the eighteenth century opened up the continental Northwest, a major contributor to timber and hydroelectric power of the region, and gateway to northern boreal forest habitats, it is now a principal constituent of the unique recreational resources of the North Shore.

*Nipigon.* The word will remain evocative: of ancient herds of woodland caribou on a thousand islands; of storied brook trout fishing (does a new record swim today in the lake's cold depths?);

and, along the northern shores of Windigo Bay and Windigo Island, of the legendary Ojibway epic of a giant of winter's ice and snow.

# SIXTEEN

# The Royal Island

It is difficult to say when an early European navigator first glimpsed the shadowed, green shores of Isle Royale, this regal island set amid Lake Superior's expanse off the coast of the North Shore. The date may have been 1618, the viewer Étienne Brulé, who was known to have brought back samples of copper. Before the Europeans, the Ojibway had visited the island for centuries. As evidenced by extensive mining of native copper, even more ancient tribes also had used the island. Isle Royale, in fact, had been one of the major sources of copper employed in intracontinental trade by prehistoric cultures.

The Ojibway had their own name for the island: *Minong*—a word meaning "copper," or "the high place," or "place of blueberries," or "island" in general, but in this case it was applied also to the island specifically. Isle Minong was the name that appeared on early French maps for many years. In 1669, the name was changed to Isle Royale, after King Louis XIV; the change reflected not only French influence, but—we may suppose—their perception of the island's unique and regal qualities.

When modern scientific explorations began in earnest in the mid-1800s, those explorers found a northern lake wilderness of incomparable beauty but with almost impenetrable borders. Dwarfed cedar, fir, and spruce lined the shorelines, their branches interlocked to create physical obstructions. And inland travel by foot was a succession of steep climbs, treacherous descents, and boggy, mosquito-filled swamps. But also inland were high ridges with sweeping views of crest and slope, bays and coves, fingerlike points of rock and small islands, and the expanse of Lake Superior on all sides.

Today, the natural character of Isle Royale remains only moderately changed. Moose, colonizing the island after the turn of the last century, now keep the formerly impenetrable vegetation more open, except on the smaller islands off the main island where these large browsers have not become established.

Little modification by human intrusion is visible, despite attempted commercial ventures over nearly the last 200 years. Commercial fishing in the waters of Lake Superior off Isle Royale's shores began around 1800; fishing continues today, but only by a handful of hardy fishermen. Copper mining was attempted in mid-nineteenth century, shortly after the Ojibway ceded their claims to the island, and continued intermittently through the latter half of the century. Logging was tried, too, although the island's thin soils never grew the stands of tall pines like the mainland's. A major forest fire swept part of the island in 1936, leaving one-fifth of it charred.

The mining, logging, and fire scars are healing as natural succession occurs, albeit slowly in places. Among the island's wild inhabitants, a life-and-death scenario proceeds as it must have done in prehistory, although now with different actors.

So the island remains today as an almost pristine example of northern wilderness, as Isle Royale National Park and a unit in the United States National Wilderness system, and as a remarkable outdoor laboratory for the scientific study of island ecology.


Even the most casual hiker on Isle Royale cannot avoid the sensation of walking continuously up and down. A lakeshore trail may be relatively even and flat, but if the hiker leaves the trail and strikes inland, it will be an up-and-down travail. The reason behind this landform is the so-called ridge-and-valley topography of Isle Royale—the unique result of volcanic events that took place a billion years ago.

During the Keweenawan age, the Midcontinent Rift System

262

*Scoville Point, Isle Royale*

opened to release the earth's molten contents to form great lava lakes. Successive flows formed many layers, separated sometimes by long intervals of time. During the intervolcanic periods, rock debris was eroded from surrounding regions, carried by wandering, braided rivers, and deposited in thick layers, which later hardened to sedimentary rock. Alternating layers of hard lava and softer sedimentary rock were thus built up. More than a dozen major lava flows, distinct in character and separated by minor flows and sedimentary layers, have been identified and named on Isle Royale.

When this area foundered to form the Superior Basin, these layers were tilted upward at the edges of the depression. We see today the result of such tilting on both sides of western Lake Superior, with lava layers slanting downward toward Lake Superior's depths. Rocks on the southeast side of Isle Royale have southern counterparts on the northwest side of Michigan's Keweenaw Peninsula, some of the same layers identified on both shores.

The ridge-and-valley character of Isle Royale results from

263

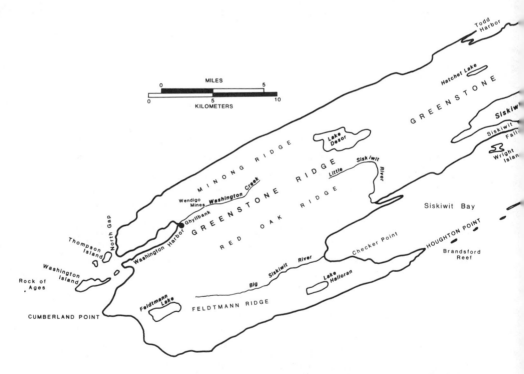

the exposed edges of these layers. The sharp upper edge of a lava layer, hard and resistant, remains as a ridge; between one ridge and the next one to the northwest—that is, the next lower and

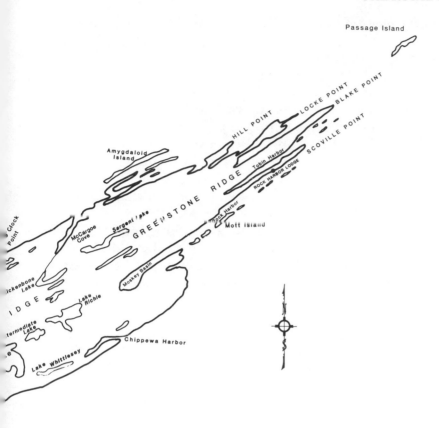

Passage Island

LOCKE POINT
BLAKE POINT
HILL POINT
SCOVILLE POINT
Amygdaloid Island
Tobin Harbor
ROCK HARBOR LODGE
Rock Harbor
GREENSTONE RIDGE
Sargent Lake
Clock Point
McCargoe Cove
Mott Island
Moskey Basin
Ickenbone Lake
I D G E
Lake Richie
termediate Lake
Chippewa Harbor
Lake Whittlesey

LAKE   SUPERIOR

therefore older layer—is a depression from which the interlayer of softer sedimentary rock has been eroded, resulting in a valley. Since the angle of tilt is only about twenty to thirty degrees—the southeast-facing slope of a ridge, or *surface* of a lava layer, is gradual; the northwest-facing slope of a ridge, the *edge* of a lava layer, is steep.

Down the center of Isle Royale runs Greenstone Ridge, named after one of the major lava flows, the backbone of the island. The Greenstone flow was the most massive, resistant, and thickest of the lava flows. Lesser ridges on either side—Minong, Red Oak, Hill Point, Scoville Point, and others—successively occur at lower elevations, the remains of the edges of lesser lava flows.

Differing from the ridges and valleys, however, is a broad area making up the southwest portion of the island. This part is underlain by sedimentary rocks of sandstone and conglomerates, deposited by braided rivers flowing eastward and carrying materials during an intervolcanic period from land to the west that is now Minnesota. The slopes to the water are more gradual; the soils are thicker and richer; and different vegetation and forest types prevail. However, within this area is the most spectacular cliff—and lookout—on Isle Royale, located on Feldtmann Ridge.

When Pleistocene glaciers melted, Isle Royale was submerged beneath icy meltwaters that filled the Superior Basin. But when the level of meltwaters fell, the jagged edges of the lava flows rose above the water's surface, and Isle Royale emerged as an entity. Around the shores of the island, the lowermost valleys were filled with water as Lake Superior subsumed the depressions and created long, narrow coves. Some of these coves now constitute ideally protected harbors, welcome refuges for ships during times of storm. Ridges stand out as narrow peninsulas and islands. Isle Royale, then, is not a single monolith, but rather an archipelago—a collection of more than 200 islands.

Inland, valley extensions form the island's lakes, ponds, and stream courses. Thirty-eight named lakes occur inland, plus about an equal number of unnamed small lakes and ponds. The largest, Siskiwit Lake, is seven miles long and 142 feet deep. There are five major streams, numerous small tributaries, most of them slow and sluggish.

The archipelago is about 50 miles long, 9 miles wide at its widest, 210 square miles in area. Shorelines are complex and total 200 miles, but including all the islands the grand total undoubtedly is many more.

Isle Royale, while politically within the state of Michigan, is geologically and climatically part of the North Shore, with a natural kinship closer to Ontario and Minnesota than to Michigan. After the American Revolutionary War, part of the international boundary beween British Canada and the fledgling United States was set to run through the Great Lakes and up the voyageurs' old trail. It was at this time, it is said, that Benjamin Franklin horse-traded the British out of the island, which was thought to be rich in copper, and he drew the boundary line north of Isle Royale.

In succession, this region of the United States then passed through a series of territories, from the Colony of Virginia to Michigan Territory. When in 1836 the State of Michigan was carved out, Isle Royale was included. The American portion of the North Shore became part of Minnesota Territory in 1849, part of the State of Minnesota in 1858. And Isle Royale, hugging the North Shore of Lake Superior, remained as part of Michigan.


The forests of Isle Royale never developed like the big woods of the mainland North Shore, with its large specimens of tall pine and dense spruce-fir woodlands. An insular environment, rocky ridges exposed to wind and storm, and thin soils did not allow the development of thick forests with large trees.

Along cool, moist shorelines white spruce and balsam fir predominate, but dry ridge-tops either contain mostly birch and aspen or have no trees. Inland valleys hold black spruce swamps. Pines are rare, restricted to certain ridge-tops and lakeshores. On the heavier soils of the southwest part, markedly different from the rest of the island, sugar maple and yellow birch form a climax forest.

The moose of Isle Royale have drastically affected the structure of the island's forests, especially along shorelines. When moose arrived on the island, balsam fir was by far the principal tree species, covering up to 80 percent of the island. But as a result of heavy browsing by moose in winter, this tree will probably never again develop to large sizes.

Browsing moose nip off twigs, effectively pruning the branches, keeping young trees at early successional stages; in this way the moose themselves help maintain their food supply. Ground hemlock, a winter favorite of moose, was once abundant, but it occurs today only on certain smaller islands where moose have not become established.

In the valleys with slow drainage and intermittent streams, beavers have had the greatest impact on vegetation. Removal of aspen and construction of dams have been the factors mainly responsible for these picturesque areas of pond and meadow.

Fire also has been an important agent affecting the forests on Isle Royale, as wildfire has in many forest areas. Among the oldest historic remains of fire are those of early miners in the mid-1800s, who burned off vegetation in some areas to expose rock surfaces for copper prospecting. There are traces of Indians having done the same.

The single fire event having the greatest impact occurred in the summer of 1936, a year of severe drought over broad regions of the midcontinent. The fire started in an area of dry slash from a commercial logging operation, near the head of Siskiwit Bay, and it burned all around Siskiwit Lake. Wind-blown sparks jumped to start new fires, burning up to the north coast of the island. Eighteen hundred fire fighters participated in the attempt to put out the blaze, but it was not extinguished for three weeks, when rain finally came. One-fifth of the island, about 27,000 acres, was left charred and black. Strict fire protection was initiated immediately thereafter. The effects of the 1936 burn are still visible, as heavy browsing by moose slows full recovery.

Logging has not been an important factor affecting Isle Royale's forests. The scanty production of large trees, as well as the isolation of the island, did not allow for the same intensive logging that occurred on the mainland. Early fishermen and miners cut trees for building lumber, probably accounting for the scarcity of white pine today, and tamarack from the valley swamps and bogs was cut for mine timbers. But only two serious attempts at commercial harvesting of timber were undertaken. In the 1890s, a Duluth logging company cut logs in the Washington Harbor area at the southwest end of the island, but a storm-caused flood washed its harvest out to Lake Superior. And a pulpwood cutting operation at the head of Siskiwit Bay in the early 1930s was destroyed in the 1936 burn that, ironically, started in the logging slash.

Another drought year, 1976, produced a potentially serious fire. Even though this fire was not suppressed, it did not spread widely, apparently because of a lack of fuel on the forest floor (possibly as a result of moose browsing on balsam fir).

Today, wildfires are not put out except where property and

life are threatened. They might be closely watched, but not extinguished. Human-caused fires, on the other hand, are controlled.

We now recognize wildfire as a natural and regenerative element in this living museum of forest succession, affecting vegetation and the wildlife it supports in a continuing sequence.

The forest mosaic is a different one now, but its dynamic character is based on the same ecologic principles that prevailed in a prehistoric wilderness.

The nugget was not your common, everyday, pea-sized lump of bright metal, lovingly fondled by a sourdough prospector. This nugget—of native copper, extracted from a prehistoric Indian pit near one of Isle Royale's historic mines—weighed nearly three tons.

It was in the Shield Archaic period, some 5,000 years ago, when the Old Copper Culture developed and mining began on Isle Royale. This copper industry roughly coincided with warmer climates; it was later to decline and disappear, perhaps as colder climates returned. Exposed at the surface in many places throughout the island, copper veins were dug out by the early inhabitants, pried and pounded with stones, and shaped into spiritual objects.

This early copper mining was no small venture; there remain on the island more than a 1,000 mining pits and holes of prehistoric origin. A modern engineer estimated that, using contemporary methods, the prehistoric digging in the McCargoe Cove area alone would have required the efforts of 10,000 men for 1,000 years.

The three-ton nugget had remained in the old pit, apparently, simply because the Copper Culture people could not move it. Now, in the nineteenth century, even greater masses of native copper came from the Keweenaw Peninsula, Isle Royale's southern counterpart, where at one time existed the most important known deposits of native copper in the world. If in the lava strata of the Keweenaw across the lake, why not also in the same formations on Isle Royale?

Mining explorations in historic times began in 1843, when the Ojibway Indians ceded these lands by treaty with the State of

Michigan. Three years later, the copper rush was well on, and four major mining companies were hard at work.

Copper mining technology was primitive in the middle of the nineteenth century but was reasonably effective for the times. Three grades of copper were recognized: mass, barrel-work, and stamp. *Mass copper* consisted of large veins and chunks weighing up to several tons; these were removed intact by prying and digging, sometimes after chiseling the larger pieces into smaller ones for handling. *Barrel-work* consisted of smaller pieces, stringy veins and sheets, that could be separated from surrounding rock, then packed into barrels for shipment, each cask weighing up to 800 pounds. *Stamp copper* consisted of small pieces bound up in rock that had to be crushed or "stamped' in order to free the copper pebbles. Several stamp mills were erected on Isle Royale, but most of the copper mined was mass or barrel-work.

Shafts were dug by hand and by blasting. They generally were less than 100 feet deep, but the maximum was 360 feet. Ore was hoisted to the surface by hand winch or, in some cases, with horsepower. All copper mined was shipped down the lakes, through Sault Ste. Marie, by sailing ship.

This early period of mining lasted only until 1855; the industry had been only marginally profitable. Later, the Civil War stimulated copper prices, and improvements in mining technology and shipping produced further enterprise. Supporting settlements developed; the town of Ghyllbank grew at the southwest end at Washington Harbor. Even a short railroad served the mines.

But real riches never appeared. Wendigo Copper Company, which had erected the most elaborate settlement and exploration system of all, went out of existence in 1892, and copper mining ceased permanently. Quiet returned to Isle Royale.

It had been a colorful era, filled with miners' hopes, some modest successes, disappointments, and some tragedies. It left a heritage of historic artifacts on the island, an integral part of Isle Royale's living museum.

With the tremendous riches of copper on the Keweenaw Peninsula across the lake, why, indeed, had there not been similar rich lodes on Isle Royale, in similar layers of lava rock? The truth is that the layers are not identical; Keweenaw's are lower, older strata. The northern extensions of the same strata now lie far under Isle Royale, beneath the waters of Lake Superior—do they, too, contain another lode of copper unequaled in the world?

Of the several economic enterprises that were attempted on Isle Royale, it was commercial fishing that began earliest, left the least permanent mark on the landscape and shores, and yet remains today, albeit at a low level of activity. Its history spans nearly two centuries.

The success of commercial fishing at Isle Royale can be attributed to two major factors. First, fishing was highly productive over the reefs, shoals, and submerged offshore lava ridges. Second was its proximity to the North Shore, which meant that hardy fishermen were available to work the boats and nets, and that nearby markets could be used to sell the catch.

The North West Company fished the waters of Isle Royale around 1800, or even before, mainly for the purpose of supplying food to its fur posts on the mainland. No detailed records of these activities remain, but some archaeological evidence suggests the location of fishing stations on the north side of the island. No doubt French voyageurs were involved in the work.

Better records are available concerning the fishing by the American Fur Company, however, beginning in 1836. By 1839, seven centers of fishing were located on the island, manned by three to five men each. These crews—again, voyageurs—made their own gill nets, fished from crude boats that were essentially oak-board barges, and packed fish in casks. These last were picked up periodically by sailing schooners for shipment to the mainland. The main establishment, at Checker Point in Siskiwit Bay, had permanent dwellings, but most of the work was seasonal, with the crews transported to summer camps for the fishing season. The American Fur Company was not long-lived, however, and by 1841 its fishing activities had ceased.

Fishing continued and expanded slowly during the middle decades of the nineteenth century. Eventually, many stations were established around the island. The most highly valued species were the lake trout and the lake whitefish, which remain favorites to this day. Both were abundant and easily taken in gill nets. A third important species was the lake herring, even more abundant, but less highly valued. A fourth fish was the oily and larger

271

lake trout subspecies, the siscowet. The siscowet was harvested mainly for its oil, which was rendered, packed, and picked up periodically for transport to the mainland.

In the late 1800s, improved technology in the form of efficient gill nets and steam-powered fishing boats resulted in a great increase in commercial fishing. Also, better methods of transport on the mainland—including railroad refrigerator cars—brought the distant markets of St. Paul, Minneapolis, and Chicago within reach. Up to sixty crews of fishermen worked seasonally during these years, and many thousands of barrels of trout and whitefish were harvested from Isle Royale's surrounding waters.

The twentieth century brought further changes. With gasoline power, major fishing companies operated from farther away—for example, Duluth—and some permanent, year-round settlements were developed on the island. Many other fishermen had their homes on the North Shore mainland or in Duluth, and some traveled seasonally to the island in their own boats to fish during the summer, returning in the fall to their mainland homes. Fishing stations surrounded the island, taking advantage of the numerous excellent harbors. Fish were abundant, and the island's fishing industry flourished.

But in midcentury disaster struck Isle Royale's fisheries, as it did throughout Lake Superior and the rest of the Great Lakes as well, with the coming of the sea lamprey. Populations of lake trout and whitefish, while saved from extirpation, were drastically reduced by the lamprey. Commercial fishing to serve distant markets essentially ceased for the two principal species, and fishermen turned to the still abundant but less valued lake herring. Isle Royale's fisheries declined greatly.

Today, only a few fishing operations remain on the island, partly as a living demonstration of the island's cultural history in the Isle Royale National Park program. Three commercial licenses remain. The catch is collected by ferryboat, which in season stops to pick up lake trout, largely to be served and consumed by visitors and residents on the island itself. The custom seems altogether appropriate—and the broiled trout a fitting regal feast.

Like its forests, Isle Royale's inland fish fauna is less diverse than the mainland North Shore's. Owing to the island's geographic isolation and its cold, inhospitable waters, many species occurring on the mainland have been unable to make the transfer. In fact, at least one new fish species and several subspecies have evolved independently on Isle Royale and occur nowhere else.

Nevertheless, some of the angler's favorite sport species do occur in the island's many lakes and few streams. Because of light fishing pressure, some populations remain high, and excellent sport fishing is available to anglers willing to work harder for it.

By far, the northern pike and yellow perch predominate in Isle Royale's lakes, as well as in bays and coves around the island's perimeter. These two species usually occur together in successful coexistence, for the perch is the pike's principal food. The relationship is a predator-prey system much like the wolf-moose relationship. The pike is most favored by the angler. It provides good fishing in about twenty-five of the island's lakes and around the island in Lake Superior, in bays and coves. The walleye, favored on the mainland, does not constitute a major sport fishery, occurring in only three lakes—Chickenbone, Siskiwit, and Whittlesey—and in Washington and Chippewa harbors.

Major streams contain native brook trout, as do three lakes: Desor, Hatchet, and Siskiwit. Although most of these are small fish, the age-old lure of the beautiful brook trout attracts many anglers. The best brook trout fishing appears to be in the Little Siskiwit River.

Anglers may be more interested in some unusual fishing opportunities around Isle Royale. One is a fishery for brook trout that live and grow in the shallow waters of Lake Superior around the island's shores and move regularly into coves and tributary streams; these are "coasters," occurring particularly in Rock, Tobin, Todd, and Washington harbors. Another special fishery is for the introduced steelhead, or migratory rainbow trout. It makes spawning runs in spring from Lake Superior, mainly at Siskiwit Falls, one of the island's few fast-water streams, where Siskiwit Lake empties into Siskiwit Bay and Lake Superior. Large rainbows are mainly fished in streams during spring runs, but also in a few of the larger harbors.

The lake trout is the major attraction for fishing at Isle Royale. Lake trout angling around the island's shores, particularly in shallow waters over reefs and flooded valleys on the southeast side, is some of the best in all the Great Lakes. When the sea lamprey decimated lake trout throughout Lake Superior, quasi-discrete populations of trout around the island were affected least, and they continue to prosper. Siskiwit Lake, large and deep, also contains this favored species.

Other species of fish inhabiting the island's streams and lakes, and in surrounding Lake Superior waters, attract little attention from anglers. These include minnows and other small species, as well as others like the suckers, which are not attractive to anglers or fish gourmets. Yet these are important in the functioning of the island's aquatic systems; they utilize different foods, some microscopic, and they themselves serve as food for larger fish, water birds, and mammals. Foxes are known to take many suckers during spring spawning runs in streams; and it seems likely that wolves at times do the same.

Some of the less common fishes are of much interest to fish ecologists and ichthyologists, isolated as they have been from their relatives on the North Shore since glacial meltwaters retreated and the island's lakes rose above Lake Superior levels.

The wildlife of Isle Royale also remains distinct from that of the mainland North Shore, partly because of management policies implemented in a United States national park, but mainly because of its insular isolation.

The number of species is lower, for one thing. The island is a microcosm within the greater system of the Northwoods; it is large enough, and the habitats are diverse enough, to provide all environmental requirements for a set of animals that live together in a state of change and reaction. And yet, as an island, its fauna is not simply the same as in the larger, human-dominated region.

Wildlife populations on Isle Royale vary in response to natural phenomena that fluctuate or, in the longer term, follow successional trends. As such, the ecosystem here, virtually unaffected now by human activities, provides one of the best examples of naturally regulated systems in North America and a unique outdoor laboratory for the scientific study of wildlife populations, in interaction with one another and with their environment.

The outstanding object of scientific research has been the

predator-prey relationship between the moose and the gray wolf, a subject which has received intensive study over many years. The value of Isle Royale for such studies lies in its insularity and its discrete borders—the shorelines—which delimit the size of the outdoor laboratory and prevent impacts from outside. The island is also the right size to enable reliable counts of animal populations.

Neither the moose nor the wolf were present in prehistoric times on Isle Royale. It was the woodland caribou that constituted the major large mammal in the old days; common up to the early 1900s, the caribou were last seen in 1925. Caribou disappeared, and moose increased, as they did on the mainland at about the same time.

Popular theory is that the moose crossed on the ice in one of those rare winters when Isle Royale was connected to the North Shore by a frozen Lake Superior. But moose are much better at swimming than they are at walking on ice; the more likely theory is that they simply swam across, perhaps about 1912-13. Moose flourished and were well established by 1915.

The wolf arrived a bit later, in the late 1940s, and was soon well established. And there began shortly afterward a series of studies on the moose-wolf relationship that has become a classic in predator-prey research.

These studies included such subjects as the techniques used by the wolf in hunting moose; social organization of wolf packs; foods, reproduction, and diseases of both moose and wolves. And—some wondered—why does the wolf howl?

Some of the important findings include the fact that the wolf hunts and kills successfully only in packs. Primarily older, diseased moose are taken, whereas young, healthy moose are virtually immune to predation. The wolf has no predator on the island, and death occurs almost exclusively among pups and old individuals.

The moose, with successful reproduction, has the capability to overbrowse its food supply, whereupon its population may decline, sometimes catastrophically. Without predation, the moose population will fluctuate drastically owing to alternating starvation and food abundance. But with wolf predation holding down moose populations, fluctuations are lessened.

Primary among these research results has been the documentation of fluctuating populations over the long term, with moose and wolves interacting in a setting of changing environmental conditions. Probably only in a large island environment, isolated from regional succession and now also from interference by human activity, could this kind of study have been effectual.

When moose arrived, they found the island exceptionally well suited to meet their environmental needs. It was an empty niche, for the caribou had declined. There were no predators, and since forest vegetation had not been previously browsed, the immediate food supply was a windfall.

So, as is usual with populations newly arrived in such favorable circumstances, moose prospered. Population numbers increased slowly at first, then rapidly in a classic form. By about 1928 or 1930, they reached a peak—estimated roughly between 1,000 and 3,000 moose for the entire island, or a density of up to about twenty moose per square mile, a concentration far greater than the island's production of food could support permanently.

Of course the population subsequently crashed. By the mid-1930s, the moose population had been decimated by starvation and disease. But a low density of only about one per square mile allowed the moose's browse to recover, aided by the 1936 fire that produced favored second growth. Moose numbers began to creep back up, close to 1,000 again by the late 1940s.

Crossing probably on ice in the cold winter of 1948-49, the gray wolf arrived from Sibley Peninsula on the mainland North Shore to invade the big island. The wolves came probably as a pack, already socially organized.

The wolf, too, found an empty niche, with abundant food: the moose. And after a slow start the wolf prospered and increased in abundance. By 1960, probably twenty wolves lived on the island.

Moose numbers dropped from the high of 1,000 to about 600, perhaps partly as a result of the sudden and new factor of wolf predation, but mainly as a result of severe winters with deep snow. At any rate, the decrease was not decimating.

In these early years of wolf-moose occupancy, the wolves were organized into mainly a single large pack, occupying most of the island as their territory. As numbers increased, more packs were organized, occupying exclusive territories that divided the island.

Over many of the past fifty years, moose and wolf numbers have fluctuated under the eyes of wildlife scientists monitoring population numbers, changes in vegetation, and other animal species associated indirectly. Ideally, the two populations existed at a ratio of about fifty moose to one wolf. But as moose increased, so then did wolves, although with a lag. When a series of severe winters came along, heavy wolf predation reduced moose, and the cycle would be repeated.

So interactions continue, still observed by biologists and park service personnel, in this outdoor island laboratory.

Another story, similar in its fluctuations but with associated species, is that of the coyote, beaver, and red fox. The coyote was originally an animal of the south and western continental prairies. But it is an adaptable animal, and at present it occupies almost the entire continent. In its dispersal north and east, it arrived in the North Shore region around 1900, and soon afterward the coyote came to Isle Royale, probably as the wolf did, across the ice from Sibley Peninsula.

This was about the same time that the moose became established, and the coyote prospered by feeding on moose carrion as well as on beavers. During the years of peak numbers and the crash in moose population, dead moose were abundant, and the coyote had a superabundance of food. Coyotes reached their maximum numbers, approximately 150, about 1949—the time when wolves arrived.

But these two wild dogs, the coyote and the wolf, closely related, were incompatible. The wolf preyed mainly on older moose, and thus little carrion was left for the coyote. By coincidence, perhaps, the beaver crashed about the same time, owing to previous high populations that had depleted its principal food tree, the aspen, around the island's streams and lakes. And so the coyote, without its former abundance of food in the form of beaver and dead moose, declined drastically. In later years, the wolf, being intolerant of another species so closely related to itself, probably killed the remaining few coyotes. The last was observed in 1958.

The red fox, on the other hand, was rare on the island in the years when the coyote was abundant. The coyote is intolerant of the fox, as the wolf is of the coyote. And when the coyote declined and then disappeared, the red fox increased in numbers. The wolf of today, being further removed in kinship from the fox, tolerates the fox such that the two populations can coexist on the island. The fox often waits at the scene of a moose kill, ignored by the feeding wolves, to glean the leavings later. Only occasionally will the wolf kill a fox, and this appears to be merely an impromptu sporting event. Furthermore, the food requirements of the two species do not conflict; other mammals such as snowshoe hares, deer mice, and red squirrels provide most of the red fox's diet.

Besides being noted for the main species in this story of succession, Isle Royale is notable for the lack of certain species. The white-tailed deer and black bear, common on the mainland, are absent. The lack of deer, carrying meningeal worm, or "moose disease," undoubtedly is a major factor in the continued success of moose on the island, although larvae of the parasite have been detected at low levels of incidence. Some older accounts report that white-tails were once introduced, but they did not survive permanently. Bears are known to prey significantly on moose calves on the mainland, and were they present on the island, moose calf production would certainly be lower. The Ruffed Grouse, esteemed upland game bird on the mainland, never was able to fly the fifteen miles of open water from Sibley Peninsula.

Other species we might think should be present—raccoon, porcupine, and woodchuck, as well as a number of mice, voles, and shrews—are also missing. Evidently, the isolation of the island has been sufficient to prevent the introduction of the complete fauna occurring on the mainland, but research has not answered all our questions.

So, why does the wolf howl? Some say the howl serves as a means to assemble the pack for a hunt or as a warning to outsiders emphasizing their territorial borders. Howling usually occurs at dusk or dawn, and there's more of it in winter. But when the pack itself howls—chorusing, with a harmonious, thrilling sound of wilderness—it seems simply a matter of socializing. They like to sing.

The ridge-and-valley topography, extended beneath the water around Isle Royale, has constituted a maritime menace throughout the history of shipping in Lake Superior.

Whereas the "valleys" have created snug harbors when filled with water, sunken "ridges" have created hidden, sharp shoals that fatally impaled many unwary or storm-tossed craft.

Early shipping to and from the island carried fish, furs, copper, and lumber; supplies to laborers and settlers; and passengers both ways. But such commerce has not existed, at least to any significant extent, for many decades. No doubt Isle Royale's reefs took their unrecorded toll of canoes, fishing boats, and early schooners, but when steam replaced oars and sail, it was the larger, more expensive, and critical human cargoes whose losses were recorded in the litany of Isle Royale's shipwrecks.

Primitive shipping to and from the island has ceased, but steamers plying Lake Superior's waters along the North Shore between Duluth, Thunder Bay, and Sault Ste. Marie have continued to ram the hidden reefs. Of about twenty-five major wrecks, for which the island's marine dangers could claim responsibity, only three were ships with business in ports on the big island; the rest were simply trying to avoid it.

Of note was the *Cumberland*, a steam-driven side-wheeler bound for Duluth that hit Rock of Ages reef in 1876, the earliest major wreck of record. Debris was washed ashore on Isle Royale's westernmost mainland point, which was accordingly named Cumberland Point. Another was the *Algoma*, a luxurious liner of the Canadian Pacific Railway. The *Algoma* was built in Scotland, and she survived a trans-Atlantic crossing safely; but in the middle of night and a fierce November gale and snowstorm, in 1885, she smashed onto the rocks of Mott Island, with loss of forty-six lives. And there was the *Bransford*, a huge steamer 434 feet long, which hit Isle Royale twice. The first time, in 1905, she was blown onto a reef off the east shore in a raging November storm, only to be lifted off again, damaged but still afloat; four years later, in a blizzard and gale, the *Bransford* went aground near the entrance to Siskiwit Bay, on what is now named Brandsford Reef (albeit with a slight misspelling). The *Kamloops*, upbound for Fort William on a stormy November night in 1927, was battered by gale-whipped seas and icing up; she simply disappeared at the northeast end of Isle Royale, near Passage Island, with a crew of twenty. The mystery remained for fifty years—until the hull of the *Kamloops* was finally found at the base of a steep submerged slope, off Twelve O'Clock Point.

Probably the most famous Isle Royale wreck was that of the steamer *America*. Owned by Booth Fisheries of Duluth, she was a ship that served Isle Royale—royally. Part-passenger, part-cargo, the *America* provided excursions in luxury style to the island. Three times weekly, the *America* made the run from Duluth, to Isle Royale, to Port Arthur; she carried various cargoes both ways, brought supplies to the islanders, and took off their fish and furs. But in the middle of the night of June 7, 1928, the *America* hit a reef in the North Gap channel and sank in shallow water, her bow remaining above the surface. All ten crewmen and thirty passengers were removed safely. She lies today, still visible with her bow in only four feet of water and stern in eighty feet, between Thompson Island and the main island. The loss of the *America* brought an end to Booth's operations in Duluth. The ship's wheel now graces the lakeside wall of the dining room in the East Bay Hotel, Grand Marais, Minnesota.

From one end of Isle Royale to the other—from Rock of Ages reef at the southwest to Passage Island at the northeast—the ancient lava ridges took their costly and sometimes sad toll.


The attractions of Isle Royale for recreation were not lost on visitors even back in the mid-1800s. And after the turn of the century, excursion boats operated to carry tourists; with the advent of small gasoline engines, private boats journeyed across the narrow waters from the North Shore to bring sport anglers, campers, and picnickers. Private tracts of land were acquired for cabins and cottages. Four resorts were soon in operation, enabling modern adventurers to enjoy this newly discovered treasure of wilderness. Soon the idea developed of setting aside the island in some kind of public park status.

The project of establishing a national park began in the early 1920s. The proposal generated some controversy, as such efforts usually do. But in twenty years the proposal became reality: in 1940 Isle Royale National Park was formally established.

Managed as a living museum in perpetuity, Isle Royale's environments, fish and bird faunas, and wildlife populations remain in the dynamic conditions of natural succession and evolution. As a national park, the island will remain a natural but changing entity, remote and regal.

Fires will burn, and forests will regenerate. Moose and wolf numbers will fluctuate with growth and reaction. Perhaps new species or forms will appear, either by immigration from other regions or by evolutionary adaptation—and some, perhaps, will disappear.

# SEVENTEEN

# A Heritage in Parks

Sprinkled around the upper crescent of Lake Superior, the North Shore's outstanding historic and environmental scenes are lodged like randomly spaced jewels on a rocky crown. They offer a fascinating and adventurous welcome to modern explorers of the region.

These sites include a score of provincial and state parks, one national monument, two national parks, two major historic restorations, and countless waysides, village parks, campgrounds, trails, and picnic grounds. They range in size from the tiniest shoreside parking space to the wildernesses of huge Lake Superior Provincial Park and Pukaskwa National Park, together taking in 110 miles of the eastern shore.

The special features of North Shore resources are widely varied. We have come to treasure greatly many of their unique values—irreplaceable archaeological remains; geological formations that teach us the physical history of our lands and waters; places where significant events in human history occurred; healthy forests adapted to local climates; fish and wildlife populations managed so that they may be used and yet perpetuated; grand scenes from hilltop or coastal cliff.

So these sites have been selected and set aside from commercial enterprise as public parks, in much the same categories: ar-

*Brighton Beach, Duluth*

chaeological, geological, historic, recreational, scenic. Recreational parks are further divided into categories of day-use, camping,
environment appreciation, wilderness.

This chapter on parks is thus organized similarly: by category of park and its objective—preservation of geological or historic site, environmental appreciation, recreation. Within each
category, all individual parks are portrayed, whether in Canada
or the United States.

The chapter is not intended to be a *guide* per se. Lacking are
many details on access, camping, trails, and other park amenities,
which are available in profusion in brochures and maps readily
obtainable from appropriate agencies in both the United States
and Canada. Beyond how to get there, and how to live, eat, and
sleep when there, however, is the greater satisfaction of understanding our heritage of history and environment in beautiful
and fascinating surroundings.

Ontario, with its much greater undeveloped land area and coastline around the North Shore of Lake Superior, has a large and diverse provincial park system. Kinds of resources vary greatly, and so do objectives. Purposes of parks range from providing picnic or swimming opportunities for crowded vacationers, to setting aside large areas for wilderness recreation and environmental appreciation. The province's parks are divided into six categories, depending on objectives, but only five of these six have been established around the North Shore.

*Historic* parks set aside locations that were significant in our cultural history; only one on the Ontario North Shore carries this designation, although most have important historical significance. Historic parks do not provide camping facilities or other amenities.

*Nature reserve* parks preserve the rare and the unique—features of geological or biological importance. Some are open for public appreciation, but the Nature Reserve category may be used also to protect fragile features for scientific study only. Like Historic parks, the Nature Reserves do not provide for camping.

The *Recreation* provincial park provides both day use and auto-available camping, in a setting that permits intensive human use. These parks normally do not encompass large areas.

Leading the list in size are the *Natural Environment* parks, which encompass large tracts with broad ecological significance and scenic qualities, allowing more extensive recreational activities like hiking, boating and fishing, and interior camping. Auto-accessed camping is usually available as well. Lake Superior Provincial Park is the outstanding example; this park is big enough to serve many objectives: day-use and wilderness recreation, historic and geologic interests, and preservation of special natural features.

*Waterway* provincial parks combine water-based recreation with history. On the north bank of the Pigeon River, Ontario is currently developing LaVerendrye Provincial Park, the only Waterway park in the North Shore drainage, commemorating the historic fur trade route.

The sixth category, *Wilderness* provincial parks, is not represented on the North Shore. The most closely associated Wilderness park is Quetico, on the border lakes.

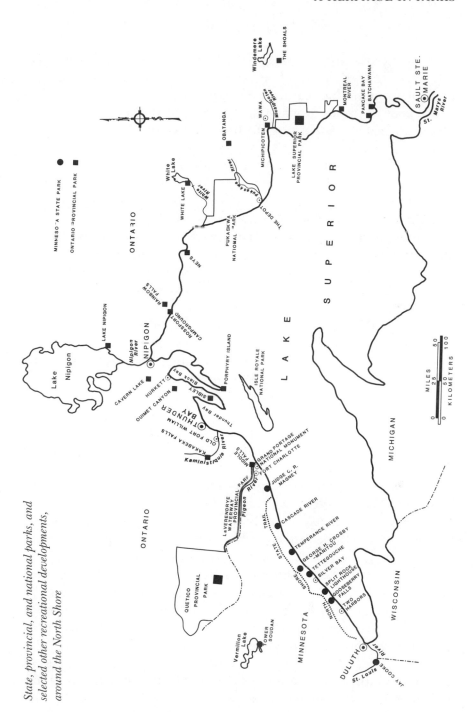

State, provincial, and national parks, and
selected other recreational developments,
around the North Shore

# Historical Parks

Most parks around the North Shore have some historical significance. Indeed, it would be rare to find some outstanding geologic or topographic feature—a large river or island, a good harbor, a coastal promontory—that had not in the past presented a utilitarian value, or obstacle, to human explorers and exploiters of an earlier day.

Nevertheless, it is not difficult, given the richness of North Shore history, to select for preservation certain special features as reminders of our forbears' endeavors, struggles, and successes in this harsh country north of Superior.

The human enterprise that left the greatest historic mark along the North Shore—at least in its romance and adventure—was the fur trade. As a natural complement to exploration, the trade was an enterprise of white Europeans that opened up the Northwest part of the North American continent. This endeavor began with French probes into the Lake Superior country in the mid-1600s and grew with the construction of quasi-military forts, missions, and trading posts, which exchanged hardware and fabric from Europe for furbearers' pelts, primarily beaver, from the natives. The fur trade reached its climax at the beginning of the nineteenth century; but two coincident events—the near extirpation of the beaver and the coming of the railroad—brought the trade to a close only a half-century later. This 200-year enterprise depended successively on three major supporting sites providing supply, organization, and staging area: Grand Portage, Fort William, and Michipicoten.

Today, a visit to these three locations can give the North Shore traveler the most complete view of the enterprise that thrived along the North Shore two centuries ago.

At **Grand Portage National Monument**, major buildings and facilities have been reconstructed in place on the clearing by Grand Portage Bay, where Sieur de la Vérendrye landed more than 250 years ago to penetrate the hinterlands. The nine-mile inland trail short-cuts a tortuous, unnavigable twenty miles of the Pigeon River and leads to the site of old Fort Charlotte, staging area for inland canoe brigades. Fort Charlotte and the trail are also included in the national monument. The heyday of the

Grand Portage entrepôt was in the last two decades of the eighteenth century, under the North West Company.

Closely related to Grand Portage National Monument on the United States side of the Pigeon River, **LaVerendrye Provincial Park** is under development on the northern Canadian side of the river. Classified as a Waterway park, LaVerendrye extends for approximately 100 miles from **Quetico Provincial Park** to **Middle Falls Provincial Park**, a corridor that includes the north bank of most of the Pigeon River and a number of border lakes. Intended to further appreciation of the exploration and fur trade history, the park memorializes Sieur de la Vérendrye, who opened the Pigeon River route to an expanding French fur trade in 1731. While emphasizing protection of this historic river trail, the park will provide high-quality recreation through low-density, back country travel. The waterway will also be designated a component in Canada's *Heritage River System*, a program of protection and managed recreation on historically significant Canadian rivers.

A half-century after La Vérendrye developed the Pigeon River route, British North West Company occupied and traded on the old canoe trails. But after the American Revolutionary War, the national ensign changed again, and the British company was forced to move up the lake to another river. At the mouth of the Kaministiquia, the company constructed a new and more elegant entrepôt—Fort William—in the first years of the 1800s.

The original location of Fort William now lies beneath industrial and residential developments in the city of Thunder Bay, but its original elegance has been recaptured in the restoration of **Old Fort William**, outside the city on another bend of the Kaministiquia River. It is administered by the Ministry of Culture and Recreation. Here the visitor can experience the spirit of the "fort" that was a state within the wilderness, serving the canoe brigades that continued to penetrate the Northwest, now by way of the Kaministiquia instead of the Pigeon.

When the old fur brigades left Fort William, upbound on the Kaministiquia for the interior, thirty miles upstream the voyageurs encountered towering Kakabeka Falls, 128 feet high. It is the highest single-drop waterfall on the North Shore—"Niagara of the North." The beauty of its amber curtains and creamy foam thundering down from hard caprock was not lost on the early explorer. But the great falls also presented a major barrier. When Frenchman Jacques de Noyon explored the Kaministiquia in the late seventeenth century, he discovered the great barrier but also a way around—Mountain Portage—a grueling carry. Today, a hydropower station often reduces water discharge upstream, and the flow over the falls may drop to a trickle. Spring and autumn are best for viewing, although the frozen falls in winter are often spectacular. A portion of the Mountain Portage track has been preserved. **Kakabeka Falls Provincial Park** is small for a Natural Environment park, but it makes a major contribution to our heritage of North Shore history.

Fort William did not last long. After the amalgamation of the North West Company and Hudson's Bay Company in 1821, the Kaministiquia River location lost its importance as an entry to the Northwest. The Hudson Bay route to midcontinent was shorter and cheaper, and after 1821 Hudson's Bay interests in Lake Superior were served exclusively by a route from James Bay to the Michipicoten River.

Historic **Michipicoten Provincial Park**, at the mouth of the Michipicoten River, has not received the degree of restoration that Grand Portage and Fort William have. But research and excavation will be encouraged in the future. The Michipicoten post served as the Hudson's Bay Company headquarters for the Lake Superior District, as Fort William fell into disuse, and functioned up to the time of the Wawa mining ventures. As railroads and steamships replaced the freighter canoes, activities at the Michipicoten location turned from furs to gold and iron ore.

At the extreme western end of Lake Superior, Minnesota's **Jay Cooke State Park** is located on the St. Louis River, headwater of the Great Lakes. The St. Louis connected Lake Superior with the upper Mississippi Valley during the fur trade, by way of the two Savanna Rivers and Savanna Portage (now also in a Minnesota state park). The St. Louis upstream from its mouth was not navigable, and the visitor today can readily tell why—while observing the wild gorge below a swinging foot bridge. The tough portage around the gorge was also termed the Grand Portage, perhaps with tongue-in-cheek, and a sector of it now renovated is incorporated in the park's hiking trail system. During times of glacial retreat, the St. Louis gorge carried meltwater from the Superior Basin; the river reversed back to Lake Superior when water levels fell, carving the gorge and sweeping valley, some 400 feet deep. In addition to campgrounds accessible by auto, the park includes some backpack campsites in the more scenic areas of the valley. Besides its historic significance in the fur trade, the park memorializes Jay Cooke, financier and railroad promoter, who acquired much land around the lower St. Louis and was largely responsible for developing the area around Duluth.

Inland from the Minnesota shore, directly north from Duluth, is **Tower Soudan State Park**, memorial to the mining enterprise in Minnesota's great iron ranges. The Soudan Mine, on the Vermilion Range, was the major underground mine for iron ore in Minnesota. Unlike the soft ores of the Mesabi, which could be shoveled from open pits, the Vermilion's ore was hard and deeply located. The mine was the product of Pennsylvania financier Charlemagne Tower, who thus left his name to a small Minnesota town, as well as to the state park. From the Soudan Mine in 1884, the first small trainload of iron ore produced by Minnesota's iron enterprise left for shipment to the docks of Agate Bay (Two Harbors) and by steamship down Lake Superior's coast. The mine closed in 1962. Trails at Tower Soudan also lead to scenes of the Vermilion "gold rush."

Symbol of the ruggedness of the Minnesota coast, and the dangers that its headlands and rocky shoals presented to shipping, mainly ore traffic, is **Split Rock Lighthouse State Park**, atop the 124-foot-high diabase cliff that appears from lakeward to be split apart in a great crack. After a number of shipwrecks along the northern coast around the turn of the last century, the lighthouse was constructed in 1910 by the predecessor of the United States Coast Guard. From that date until 1968, the great light warned ship captains of the dangerous waters below the cliff and helped establish their bearings. After that, modern electronic navigation aids took over the job of guiding ships along this coast. A new visitors' center displays much of the maritime history along the North Shore.

Two provincial parks set aside areas that were important in North Shore timber harvest. Inland from the eastern shore is **The Shoals Provincial Park**. Although classed as a Natural Environment park, it provides access, by canoe, to a historic lake and river area that witnessed the most extensive logging operation in the region of the eastern shore: the Austin Nicholson Company's harvest of jack pine for railroad ties. During the period 1901-31, the company produced millions of ties, cutting jack pine from the lands around Windemere Lake and other nearby lakes and rivers, including Prairie Bee Lake where part of The Shoals park is located. From the access on Prairie Bee, a canoe route leads north through Lower Prairie Bee Lake and connecting waterways to Windemere Lake. Here is the Canadian Pacific Railway siding, ghost town of Nicholson, and the old mill site where were cut the jack pine ties that carried Canadian Pacific trains all around the North Shore. Much of the park surrounds Little Wawa Lake, on which are located the main campgrounds. But many backcountry campsites are located on canoe trails, including the route to Windemere.

**White Lake Provincial Park**, also a Natural Environment park, is located in another area significant to the history of logging the North Shore. One of the largest pulpwood logging enterprises was the Abitibi Power and Paper Company's operation around White Lake and its upper tributary, the Shabotik River, which removed immense quantities of black spruce from this area. In the 1940s and 1950s, Abitibi rafted spruce logs down the lake, through White Lake Narrows, and on down the White River to Lake Superior, where the pulp logs were rafted again. The last drive was in 1964.

The park is located along the west side of White Lake, between the Narrows and the head of the White River, and includes some remains of the pulpwood industry. An established canoe route of some thirty miles loops around through a half-dozen lakes, with brook trout, walleye, and northern pike fishing. A longer canoe route starts to the east in Negwazu Lake and comes down the White River some seventy miles, through the town of White River and to White Lake; in this headwater reach, brook trout are the chief sport fish. From the provincial park, a white-water canoe trail descends to Lake Superior, another forty-five miles. Many primitive campsites have been established along the way.

# Nature Reserves

Most prominent of the Nature Reserves on the North Shore is **Ouimet Canyon Provincial Park**, where a thick Precambrian sill of diabase has opened in an awesome, vertical-walled trench 500 feet wide and more than 300 feet deep. The sill, over a billion years in age, has been eroded by repeated freezing, thawing, and water flow; it is thought that the canyon was a major route for exiting meltwater from the Nipigon Basin to the north, during times of glacial retreat. The rock-strewn floor of the canyon is so deep and shaded that the cold of winter rarely completely dissipates, and consequently a cold microclimate harbors a variety of disjunct Arctic plants, relics from the Ice Age. The site is available only to permitted scientific study.

Other Nature Reserve parks, not yet developed for public use, include **Montreal River Provincial Park**, on the eastern coast, with old cobble beaches, resulting from successive glacial lake levels in the Superior basin; **Cavern Lake Provincial Park**, another canyon in diabase sill, with a deep, Precambrian cave and disjunct Arctic plants, west of Nipigon; and **Porphyry Island Provincial Park**, at the tip of Black Bay Peninsula, representing the geology of Keweenawan age and insular plant associations.

# Recreational Parks

Ontario's parks in the Recreation category serve objectives of day-use picnicking, vehicle-camping, and hiking, sometimes with swimming beach. They are always located near some natural feature of scenic or geologic interest. North of Sault Ste. Marie are two small provincial parks readily accessible to the traveler on the way around the North Shore. **Batchawana Provincial Park** is a day-use-only park, with a swimming beach on Lake Superior, and **Pancake Bay Provincial Park** includes a vehicle-camping area with intensive use, as well as trails and an exceptionally large sand beach that once accommodated fur traders as a stopping point on their way west.

Near the town of Rossport on the northern coast is **Rainbow Falls Provincial Park**, a Recreation park that combines inland camping, scenic waterfalls, and Lake Superior camping. Located

on Whitesand Lake, the park includes campground, day-use picnicking, hiking trails, and boat access for fishing. At the outlet of
Whitesand Lake, the Hewitson River drops over fractured granite
in delightful Rainbow Falls and several smaller cascades. Nearby
**Rossport Campground**, associated with the provincial park, is located on Lake Superior and provides camping and a beach along
the big lake.

Farther west, **Middle Falls Provincial Park** provides camp
and picnic grounds along the Pigeon River, and several miles of
hiking trail along rapids and cascades. The main falls, about a
twenty-foot drop, is one of the reasons for the Grand Portage,
which for seventy-five years served traders and voyageurs in their
circumvention of the lower twenty miles of rugged river. From
the vicinity of the park, trails lead three miles downstream to
High Falls, and upstream several more miles to cascades and
gorge.

Farther yet up the Pigeon, above the Grand Portage trail, is
Partridge Falls, the last major obstacle to the voyageurs of old after leaving Fort Charlotte. No grand scene here, because both
river and falls are hidden in lush forest, but the misted, cool
gorge and lacy falls of seventy feet are impressive. Partridge Falls
is accessible by a forest road through Minnesota's Grand Portage
Indian Reservation. A portage path around the falls (legacy from
the fur trade) permits a canoe trip down the river of about three
miles to Fort Charlotte, at the upper end of the Grand Portage.


Among parks intended for intensive recre-
ation are a group in Minnesota that can best be described as "waterfall parks"—those locations containing the scenic attraction of
falling water and misty canyons—of which the North Shore is so
abundantly endowed. The fascination of the force and sound in
waterfalls—even the rumble of unseen cataracts below ground—
seems universal.

The diverse, tilted lava formations along the Minnesota
coast, the North Shore Volcanic Group, are particularly suited to
the production of waterfalls, and most of Minnesota's North
Shore parks are consequently located near falls.

Not far west from the border Pigeon River is **Judge C. R.
Magney State Park** on the Brule (or Arrowhead) River. A longer

reach of river is available here for spawning steelheads, which are not blocked in their spawning run for about a mile, and fishing in both spring and fall is often superb. The major waterfall attraction here is the "Devil's Kettle," reached a mile or better upstream by trail. Here the river splits at the rim of its main fall, and while one half falls in steps seventy feet to the bottom of a narrow, lava-lined gorge, the other half disappears into a stream-bottom kettle—a round hole ground in water currents ages ago by pebbles and stones—apparently to join the main river again farther downstream. Legend, however, claims a mysterious and unknown destination. The park is named for Clarence R. Magney, former Duluth mayor and Minnesota supreme court justice, who early in this century devoted much personal effort to the protection of North Shore waterfalls and establishment of Minnesota's state park system.

Next westward along the shore is **Cascade River State Park**, named appropriately for its main river feature—a long series of low falls and cascades that tumble down hundreds of feet through a deep, misted, fern- and cedar-filled gorge. A visitor trail climbs up one side, crosses the gorge on a high footbridge, and winds back down. Although the stream near its mouth is short for ascending spawning fish, it is one of the most popular and productive trout and salmon areas. The stream mouth and gorge, while apparent focal points of the park, do not constitute all of it, however; twelve miles of scenic coastline are also included.

**Temperance River State Park** is one of the most unusual of the waterfall parks, in that much of the reach with falls is underground. In deep, narrow, colubrine gorges the river rumbles through gashes in the lava rock. Spalling of the lava on the sides of the gorge leaves almost vertical walls, and much of the river course cannot be seen from the top. Downstream near the mouth, the lower gorge and estuary offer some of the best North Shore steelhead and salmon angling.

Most popular of the waterfall parks is **Gooseberry Falls State Park**, on U.S. Highway 61 and closest to Duluth. Several large falls and smaller cascades bracket the highway, but recreation sites are located along the rocky Lake Superior coast.

Many other waterfalls and scenic river reaches have been selected for locating waysides, picnic grounds, overviews, trailheads, and fishing accesses. Some are viewed best from a highway bridge, and some are reached only by an unmarked trail along streamside. Among the more notable are the falls of the Beaver River at Beaver Bay; Caribou River, with a magnificent high falls about a half-mile by trail upstream from the highway; Cross River Wayside with a wide, lacy falls by the highway and a long, cascading reach in a gorge farther upstream; good trout and salmon fishing in the Devil Track River and Kadunce Creek, with deep, narrow canyons and blocking waterfalls upstream; the tumbling Flute Reed River near Hovland and the border.

The **North Shore State Trail**, on the Minnesota shore, crosses many streams. Over 150 miles long, the trail offers summer hiking, backpacking, and horseback riding on seventy-five miles, and winter ski touring and snowmobiling for its full length. Located about ten miles inland, the trail avoids parks, towns, and crowds along the shore.

Perhaps the premier waterfalls in terms of its spectacular scene is High Falls on the Pigeon, border river, located about one mile upstream from the highway bridge and customs offices. Neither Minnesota nor Ontario have developed a park to include these falls, but primitive trails lead to the falls on the Canadian side.

# Natural Environment Parks

Provincial parks in the Natural Environment category along the Ontario shore are the largest and offer the visitor the widest views of Lake Superior country. Diversity is high. From simple day use for picnicking or swimming to a backcountry hiking or canoe trip, scenes of river and waterfall, rugged coast and remote beaches, large lakes or hidden trout brooks, are seemingly endless in their variety.

**Lake Superior Provincial Park** is the largest of these, lying along the eastern shore from near Wawa on the north to the Montreal River on the south—sixty miles of shoreline.

These parklands are extensive enough to constitute a major segment of the North Shore, with rocky headlands, forested mountain slopes, protected harbors, raised cobble beaches, sandy swimming beaches, river mouths below waterfalls and tumbling

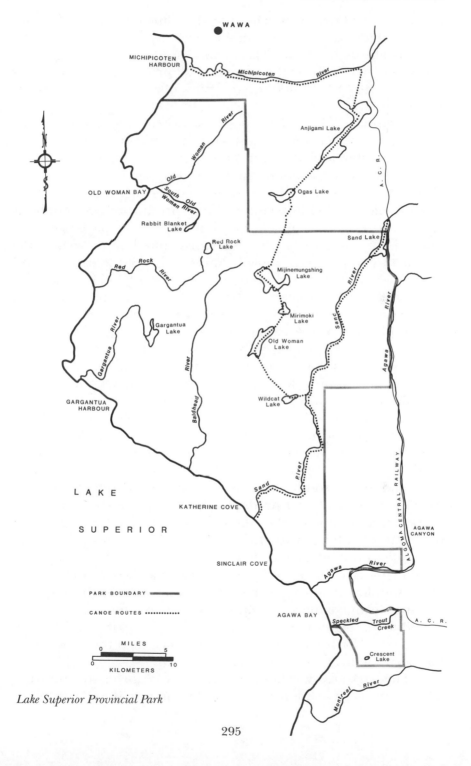

*Lake Superior Provincial Park*

cascades. The highway touches the shoreline only rarely. The interior includes large tracts without roads, trails, or rivers, but the interior also holds many lakes, canoe routes, and hiking trails, too. This park is by far the largest of all provincial or state parks on the North Shore—600 square miles of unspoiled Lake Superior country.

Beginning with Louis Agassiz's visit in 1848, the eastern shore early became a major attraction for tourists. It was accessible only by ship at first, around the turn of the century: the *Caribou* and the *Manitou*, out of the Sault, dropped fishermen at several points along the shore, including the lighthouse at Gargantua Harbour. Other visitors built private lodges simply as retreats to enjoy the northern wilderness. The park area was opened up considerably to fishermen and hunters when the Algoma Central Railway (now forming much of the eastern park boundary) was completed in the region in 1912; lodge accommodations were made available at Sand Lake, Old Woman Lake, and Agawa Canyon. In the railway's early years, some of the Group of Seven, painters of the eastern shore, traveled the interior in a special boxcar that served as both home and studio, to produce some of our best-loved canvases of the Lake Superior country. The railway still provides a major access to the park's interior.

Recognized for its high quality and diverse recreational values, Lake Superior Provincial Park was officially established in 1944. By that time, the highway had reached from Sault St. Marie northward to the Montreal River, at the south end of the park. In its first period of fifteen years, the park served visitors by rail, boat, and, to the south end only, by car. Private lodges provided for visitors' recreational needs, mainly along the shore, and the more intrepid came by the Algoma Central to put in canoes for backcountry fishing trips. In these early days, the train would stop to let off anglers and their gear wherever a lake or stream beckoned—as well as to pick up or let off mail, ice, groceries, and other supplies. A waving arm on the track was enough to bring a huffing steam engine to a halt, for service to a tourist.

But the highway was completed into the park in 1959 (all around the North Shore in 1960), and since then the automobile has been the main access mode.

The park area was home to the Ojibway for many centuries. Then European explorers passed by, and traders from both

North West Company and Hudson's Bay Company occupied fur posts, for example, at the mouth of the Agawa River. When the Algoma Central Railway went through, pine and pulp logs were stripped from much of the interior and shipped south by rails—Mijinemungshing Lake being one of the major logging areas. Commercial fishing for lake trout flourished along the coast in the early decades of the twentieth century, up to about 1950 and the sea lamprey depredation; Gargantua Harbour held the major fishing station along this part of the coast.

Today, the old fur posts are gone, along with the logging camps and fishing stations. Some modern logging operations are still permitted in the interior, and lake trout are coming back. In place of the bustle of commercial enterprise, however, is a new spirit—this time of resource appreciation and enjoyment of the natural environment.

Lake Superior park is almost big enough to be all things to all people. Management of the park is through zones, similar to the provincial park classification itself: Wilderness, Natural Environment, Nature Reserve, Historical, Recreation-Utilization. Within this scheme are three auto-accessible campgrounds (Crescent Lake, Agawa Bay, Rabbit Blanket Lake); six day-use areas with picnic grounds (three in the campgrounds, three on Lake Superior—(Sand River, Katherine Cove, Old Woman Bay); a boat harbor with federal dock (Sinclair Cove); seven major canoe routes totaling more than 100 miles; more than fifty miles of established hiking trails with backcountry campsites.

The Wilderness zone is managed in two parts: coastal, for shoreline features on a hiking trail, between Gargantua Harbour and Old Woman Bay, and the interior zone, accessed mainly by road to "Mijin Lake."

The Natural Environment zone—comprising fifteen designated areas—is for low-intensity recreational use, mostly hiking and canoeing. These areas include broad landscapes in high uplands, remote forested areas, rivers and gorges, and the many bays, headlands, beaches, and islands along the coast.

Nature Reserve zones are small in area but protect special geological and forest types, such as distinct rock faults and rare sedimentary rock features, glacial eskers and cobble beaches, bogs, tracts of sugar maple, and disjunct Arctic flora. Historical zones include prehistoric occupation sites, fur post, fishing settlements.

Recreation-Utilization zones include both low-intensity recreational activities and some commercial harvesting of timber resources. Recreation activities include hiking and backcountry camping, canoeing and fishing.

Fishing in the park centers mainly on brook trout in interior lakes and streams, although walleyes and northern pike can add to the angler's catch in some lakes. The park is big enough, too, that some hunting is permitted, for moose, grouse, and snowshoe hares, in the interior east of the highway. Much of the fishing is still accessed by rails of the Algoma Central Railway along the east boundary. Most of the park is in the Great Lakes-St. Lawrence mixed forest, with the boreal forest extending down a little into northern parts.

Few roads have been built in the park; other than the main highway (No. 17) passing north-south through the park, secondary roads lead to Frater Station on the Algoma Central Railway, to Gargantua Harbour and the coastal trail, and the Mijinemungshing Lake road which accesses some of the canoe route network. The greater part of Lake Superior park remains open to personal exploration, by foot and paddle, of an expansive scene of forest, river, and wild coast.

**Obatanga Provincial Park**, another large Natural Environment park, lies inland from the coast, crossed by Highway 17 as the road swings around the Pukaskwa Peninsula. Although thus easily available by highway, most of Obatanga is remote and quiet. The park contains thirty-two lakes, large and small; a five-day canoe trip through some of them and their connecting rivers provides an experience exemplary of the inland lakes of the North Shore. Good fishing can be expected along the canoe route, mostly for walleyes and northern pike. But the main quality of Obatanga is its gateway to an outstanding example of the interior lake country.

**Neys Provincial Park** comprises most of the Coldwell Peninsula, projecting southward from Lake Superior's northern coast. Roadless, the interior of the peninsula remains largely untrod. Here granite ranges rise steeply to windswept summits 900 feet above Lake Superior, overlooking small lakes, trout brooks, and beaver ponds. Vegetative cover varies from nonexistent on exposed rock hills, to deciduous woodland of birch and aspen on central slopes where past fires once burned, to the lowland luxurience of old boreal forest of spruce and balsam. In a harsh microclimate, exposed shores hold examples of disjunct Arctic-alpine plants. Here in this wild peninsula and on some offshore islands still live some woodland caribou. Neys Provincial Park is the Natural Environment park that perhaps best exemplifies a wild and rugged northern coastline.

Neys is relatively small, at 8,150 acres, for a Natural Environment park. But since almost all of it juts into Lake Superior, the park includes some of the longest and wildest shore, exposed to some of Superior's wildest storms.

Neys boasts its share of historic scenes. The peninsula is bordered on the west by the Little Pic River, which in its day carried great pulpwood drives; on the east is Coldwell Harbor and the remains of one of the North Shore's most productive fisheries; in the developed area of camp and picnic grounds was Neys Prisoner of War Camp, which held German prisoners during the Second World War.

And from some unknown prehistoric human culture, on the most remote southern shores of the peninsula and some offshore islands, on raised cobble beaches, remnants from glacial lake levels, are "Pukaskwa pits"—cobble-lined depresssions that perhaps served early peoples with some religious significance, as hunting blinds, or storage of frozen or iced fish.

Neys Provincial Park is still under development, but it will not be developed for intensive use. In plans for the future are construction of some interior trails and primitive backcountry campsites. Possibly, some offshore islands, primarily large Pic Island, will be added to the park, along with the islands' caribou and Puckasaw pits. The emphasis will be on a wilderness experience on some of the most rugged of Lake Superior's unspoiled shores.

**Lake Nipigon Provincial Park** (formerly **Blacksand**) is located on the southeast corner of Lake Nipigon, the most northern of the Lake Superior parks. Nipigon, discussed more thoroughly in an earlier chapter, provides access to fishing in the lake and, for a day's river trip, through the valley of the Nipigon

River. Smaller than most Natural Environment parks at 3,600 acres, Nipigon provides some unusual forest appreciation opportunities, as well as lakeshore recreation, archaeological and historic sites, and trails to a spectacular view of the big island-studded lake from the edge of towering diabase cliffs.

**Sibley Provincial Park** lies astraddle the high mesas of diabase caprock on Sibley Peninsula, stretching twenty-five miles southward from the North Shore into Lake Superior. On the west side, diabase cliffs rise to towering heights, some vertically to 450 feet, and overlook Thunder Bay; sweeping views from the trails are spectacular. Beyond, farther west, lies the city of Thunder Bay.

Sibley is a large Natural Environment park, nearly 100 square miles. Large tracts of back country make up most of the park, yet the wild boreal forest and mountainous shores are readily accessible by foot trails. On the peninsula are more than twenty lakes.

Viewed from the city of Thunder Bay, the southern segments of the peninsula lie on the water's horizon in the shape of a giant, supine Indian, asleep on the waters of Lake Superior. Diabase sills make up the formation; blocklike mesas form its features of head, folded arms upon the chest, and outstretched legs. Indian legendry assures us that it is the stone manifestation of Nanabozho, protector of the Ojibway, who smote two white men seeking silver at the tip of the peninsula, with one of Lake Superior's most violent storms. The silver, of course, was later found anyway, on the tiny rock near the peninsula's tip soon named Silver Islet, a storm-battered outcrop with rich veins of silver ore and a mine of historic fame.

Although Minnesota has not developed a park classification system like Ontario's, two state parks along the northwestern coast meet the criteria of the Natural Environment category.

**George H. Crosby-Manitou State Park** is located along the last approximately ten miles of the Manitou River, where it drops 800 feet in its tumble to Lake Superior. In the middle of this reach, the river drops about 250 feet in one mile through a jumbled, rock-walled gorge known as the Cascades. The stream for several miles is almost a continuous succession of waterfalls, cascades, and rapids. Pocket water within the rapids and intermittent deep, black pools support one of the best brook trout populations on the North Shore.

Scattered along the river's tumultuous course are backcountry campsites, accessible by hiking only. The park also includes a small lake—Bensen Lake, designated trout water—with additional backpack campsites. Crosby-Manitou is one of Minnesota's largest, at nearly 5,000 acres. It represents a recent trend toward setting aside more North Shore resources for appreciation in a remote wildland setting. The park owes its existence and its name to George Crosby, a mining entrepreneur on the Mesabi and Cuyuna iron ranges, who donated a large tract of forest for the park's establishment.

**Tettegouche State Park**, one of Minnesota's newest, includes a great diversity of natural features. At 4,800 acres, it is also one of the state's largest North Shore parks.

Tettegouche takes in about one mile of Lake Superior shore (formerly Baptism River State Park), including the mouth of the Baptism River and Shovel Point, a lava headland protruding into Lake Superior above sheer cliffs. From trails out on the point, a splendid view is to be had down the shoreline, including the mouth of the Baptism, grottoes in the cliff, and tilted lava layers. Westward is Palisade Head, a massive lava headland with the highest vertical cliff on the Minnesota shore, over 200 feet. Upstream on the Baptism, other trails lead to falls and cascades, including Minnesota's highest falls, seventy feet (except for High Falls on the border Pigeon River).

But the Baptism River and Lake Superior cliffs are only a small part of Tettegouche. Inland away from the river lie remote tracts of high forest with wooded slopes, deep valleys, small streams, and four lakes. Except for the highway at the mouth of the Baptism, no roads cross the park; but hiking trails lace the area, touch all four lakes and several high overviews. Tettegouche is a hiking and climbing retreat, with big deciduous woods and sweeping hills. Maple woodlands in autumn are a pageant of color.

Fishing in the four interior lakes is by foot only and consequently difficult. More popular are steelhead and salmon fishing, in season, in the Baptism near its mouth, some of the North Shore's best.

In the mid-1970s, much of what is now Tettegouche was sought for a taconite tailings dump by Reserve Mining Company of Silver Bay. The area included the valley of Palisade Creek and one of the high lake basins. After viewing this spectacular high country, however, a United States District Court judge in 1975 ruled against the tailings dump, and the high valley was saved, later to become part of Tettegouche State Park.

# Municipal and Local Parks

All around the shore, numerous municipal parks, historical waysides and museums, picnic grounds, and beaches with special features offer day-use recreation, exploration, and occasional swimming. A random sampling must suffice.

At the eastern end, near Sault Ste. Marie, is Hiawatha Park, with the falls of Crystal Creek as it drops from the 500-foot-high Gros Cap granite escarpment, to empty into the St. Marys River; north along the shore, the Harmony (or Chippewa) River drops over more falls in Chippewa Falls Park to empty into Harmony Bay with its sweeping swimming beach; just south of the Montreal River is Alona Bay roadside park, near the first Canadian discovery of uranium. In the Michipicoten-Wawa area, numerous parks and historic sites include the old mission, Hudson's Bay Company fur post, a stretch of wild sand dunes, spectacular falls on the Magpie River, the old Algoma Central Railway harbor facilities, and a trailer park on Wawa Lake commemorating the 1897 William Teddy gold finds.

The Lakehead abounds with historic and recreational sites: the excellent Nipigon Museum; Hurkett wildlife area, noted for birding; the Thunder Bay area with Trowbridge Falls and Current River parks that include unusual waterfalls, wild river stretches, and camping, on the Current River; Mount McKay, a diabase mesa 1,000 feet above Lake Superior with sweeping views of the city of Thunder Bay, the Kaministiquia River delta, the Sleeping Giant, and the bay's islands.

On the western end of the shore, Brighton Beach, at the edge of Duluth, includes little of a beach but an extensive strip of picnic grounds along the lake, where salmon fishing from the rocky shore is often successful. The town of Two Harbors includes historic ore docks, two preserved locomotives from the iron mining era, and depot museum on Agate Bay, and on Burlington Bay municipal camp and picnic grounds. Up the lake a short way from Two Harbors is Flood Bay wayside, with picnic grounds along an outstanding flat-pebble beach noted for agate-hunting and beach-combing. In Grand Marais harbor, the tombolo (a rocky island connected to the mainland by current-deposited gravel) that separates East Bay from the main harbor, is known as Artists Point, a wild, wave-battered rocky shore.

# National Parks

Lying twenty miles off the coast near Grand Portage is Isle Royale, actually a fifty-mile-long archipelago of some 200 islands. Most of them are small and surround the main island. Formed of the upturned edges of Precambrian lava layers, the archipelago projects above Lake Superior's surface because of a massive fault in the ancient rock, displacing it from the mainland. On most days, the big island is visible from the mainland shore between Grand Portage and Thunder Bay. But there frequently comes a day with one of Superior's common fogs—and the archipelago becomes detached and invisible, lost from its mother shore.

The splendors of Isle Royale were amply apparent to visitors in the early years of this century. Isle Royale had been used even earlier for commercial purposes of copper mining, logging, and fishing—and in prehistory by native peoples, for copper, some 5,000 years ago.

But when the copper and pine logs ran out, and when water transportation developed to the point that private persons could venture to the island simply for enjoyment of its natural attractions, a great stirring began for setting aside this magnificent bit of displaced North Shore as a park.

The physical description of the main island, its cultural history, and the unique ecological evolution of its wildlife associations have been set forth in an earlier chapter. The establishment of **Isle Royale National Park**, in 1940, meant that these physical features and the dynamic wildlife community would remain relatively undisturbed as a living museum. Even a bit of cultural history—commercial fishing—lives on, conducted in the old pattern.

Today the island remains one of few insular national parks, without automobile access but nevertheless readily accessible by a choice of other travel modes.

The island's many delights are mainly experienced by hiking or canoe and boat. More than thirty campgrounds have been established, available by such personal modes. Transport by ferry to nearshore camps is another choice. Linking the campgrounds are coastal canoe and boat routes, winding among smaller islands and bays, and 200 miles of hiking trails that trace high ridges making up the backbone of the big island. Rock Harbor Lodge, with overnight accommodations, meals, and boat rentals, is located near the northeastern end.

Management of Isle Royale National Park (a component of the United States Wilderness Preservation System) is planned on an uncrowded basis. Regionally, intensive use and motor-accessible camping are provided elsewhere—for example in the mainland's national forest and state parks—preserving the island's intrinsic values as a wilderness ecosystem.

In the far northeastern corner of Lake Superior, the Pukaskwa peninsula extends into the lake, responsible for a significant part of Lake Superior's configuration. This peninsula does not jut sharply out from shore, nor make a stabbing thrust into the great lake; in gross outline it is subtly rounded, hardly noticeable with a quick glance at Lake Superior's map. It is the largest area along the entire North Shore that is roadless—highway and railway sweep far inland—and the peninsula therefore remains largely remote and wild. This part of the shore—harsh and stormy, rock-bound, unruly—remains also the longest reach of North Shore coast untouched by road or rail.

Pukaskwa (say "Puckasaw") was named after one of its major rivers, which in turn has the source of its name in Indian legend. Carved out of the peninsula and comprising some of the region's most scenic coastal and riverine landscapes is **Pukaskwa National Park**—nearly 725 square miles in area, embracing 50 miles of shoreline.

Officially opened in 1983, Pukaskwa is a new addition to

Parks Canada, the system of Canadian national parks. It provides a broad diversity of recreation—from intensive day use to demanding wilderness experience—along some of Superior's wildest coast. Within these broad recreational objectives, the park also provides, by reason of its great expanse, the protection of isolation to some fragile biological and archaeological features.

Pukaskwa National Park is divided into zones that are similar to the classification used for provincial parks. The zones include Special Preservation—for a remnant but healthy caribou herd, an Arctic-alpine flora that flourishes on the peninsula's ice- and storm-battered coast, and ancient Pukaskwa Pits on cobble beaches. The Wilderness zone (most of the park area) includes dunes, coastal hills, interior plateaus, and most of the major river basins, to be enjoyed only by hiking, canoeing, and skiing, but with no motorized access. Natural Environment zones are for widely dispersed recreation: hiking, picnicking, swimming, fishing. And Outdoor Recreation areas provide auto-accessible campgrounds, boat launch, beach, short trails. Some inland lakes, with motors allowed, are also classified in the Outdoor Recreation zone.

In all zones, the visitor is aware of the wildness of this part of Lake Superior's northern coast and, inland, the remoteness of boreal forest.

Many miles of hiking trail are planned, including trails on the coast, interior highlands, lakes, and rivers. The Coastal Hiking Trail follows the shore, with its almost ever-present view of Lake Superior, for the length of the park's coastal perimeter, fifty miles, down to the mouth of the Pukaskwa River. Along the trail, shores rise to 1,000 feet above Lake Superior. Near the park headquarters at the northern end of the park, in a day-use Natural Environment area, is the shorter Southern Headland Trail, providing a spectacular introduction to the wild character of this part of the shore.

River canoeing is available on the wild White River, down from White Lake Provincial Park to Lake Superior, and on the Pukaskwa River (in early spring, before June, when its water levels are up). Another, interior canoe trip connects several inland lakes, with good lake fishing.

The national park's main canoeing attraction—unique as an established route along the North Shore—is the Coastal Canoe

305

Trip. Along Lake Superior's unpredictable littoral waters, this canoe trail winds in among nearshore islands and coves, along cliffs and headlands, beaches and gravel bars at river mouths. From park headquarters, the route covers the fifty miles of park shore to the Pukaskwa River, a canoe alternative to the Coastal Hiking Trail. Continuation by canoe to Michipicoten Harbour would entail about an additional equal distance. For the entire distance, Pic River to Michipicoten Harbour, canoeists should set aside up to ten days, which allows time for lingering, fishing, and windbound layovers. The coastal strip and islands in the park are zoned Wilderness areas, with many Special Preservation sites located along it.

Pukaskwa is a carefully planned park that provides both dayuse recreation and backcountry experience, emphasizing the natural character of one of the few remaining long and remote stretches of wild North Shore environments.

In a search for a national park within its borders, the state of Minnesota in the 1950s considered, among other choices, its portion of the North Shore as a prime candidate. But by this time, U.S. Highway 61 followed the shore all the way from Duluth to the Pigeon River, almost within sight of Lake Superior; and along with the highway came the inevitable services of food, lodging, and amusements to cater to the tourist. Traffic was given a further boost when Ontario's Highway 17 was completed—to permit automobile travel all around the lake—in 1960.

Furthermore, in the 1940s, permits had been issued to Reserve Mining Company for its taconite-processing plant at Silver Bay, a huge facility from which enormous quantities of taconite tailings polluted Lake Superior.

So it was that the unspoiled, beautiful Kabetogama Peninsula in Rainy Lake, on the border, won out. The establishment of Voyageurs National Park in 1975 brought Minnesota its national park. A large, wild area, like Kabetogama, simply was no longer present along the Minnesota North Shore.

In retrospect, it may well be questioned whether, in view of geographical practicalities, it would ever have been politically possible, or even desirable, to establish a national park within the policies of the park service along this part of the coast. It is interesting, anyway, to consider what the Minnesota coast would have been like, had steps been taken a century ago to preserve in permanent public ownership these enchanting Superior shorelines.

# Special Places

The time is long past when Lake Superior's waters washed upon a northern coastline of wilderness. On the day Étienne Brulé and his company of adventurers first set foot upon these shores, the pristine frontier began a retreat. On maps of the early Jesuit explorers, the white spaces were filled in. The unknown became known. And when permanent white settlements expanded only a little more than 100 years ago, after the fevers of fur trapping, mining, and logging were spent, the loss of the pristine became irrevocable.

And yet, in many special places, the loss of wilderness has not destroyed *wildness*—that quality of harmony between land and the forces of nature controlling the land's character. Nor has the land lost its potency for renewal of the beauty of forests and rivers and shorelines.

Most of the actual shoreline—that is, more than half of its approximately 1,000 miles—exists today in the same primeval condition viewed by the first French explorer. In itself, this is an amazing circumstance in the Great Lakes region. Other parts of the shoreline, once severely altered by human activity, have now reverted to wildness. Most of the forested lands within the northern Lake Superior drainage—that is, much more than half—have

308

been modified drastically. But even here, wildness exists in portions, some tiny, some larger.

And throughout these woodlands and along the coast many thoughtful, dedicated persons in two countries work for the preservation of wildland values in public forests, reserves, and parks. These are values in special places that, renewed through changing seasons and centuries, renew the human spirit as well.

Sadly, there are some portions of the North Shore that have been badly mistreated. A few years ago, Reserve Mining Company, producing taconite pellets in its plant at Silver Bay, Minnesota, discharged its tailings wastes directly into Lake Superior, and an expanding delta of tailings grew like a spreading cancer into once crystal shoals. This pollution of Lake Superior's waters generated enough public reaction that the company was forced by court order to dispose of its wastes on land, and although a steam-spewing scar remains on the shore, the malignancy is checked for now. Superior's waters downcurrent from Silver Bay are clearer.

The pulpwood and paper industry, so financially enriching in the public economy around the North Shore, was at the same time responsible for ejecting paper-mill wastes containing sulfurous and mercuric poisons into Nipigon Bay, the St. Louis River estuary, and other places along the shore. Now, modern technologies that have been directed toward cleaner effluents have cleansed these North Shore waters of most of the poisons.

In secrecy, the United States Army Corps of Engineers dumped 1,450 barrels of military waste in deep water along the Minnesota shore, in late 1959 and early 1960. Although searches have been made, the barrels have never been found, nor is it firmly known what they contained.

Now, it seems unlikely that permits will ever be issued again for taconite wastes to be dumped into the lake, nor for unchecked manufacturing wastes. Nor is it likely that a governmental agency of either the United States or Canada, military or otherwise, will again attempt a clandestine disposal of toxic waste.

I do not believe that new industrial enterprise will threaten North Shore waters in the future. Not true ten years ago, new resolve in both countries will now prevent the disasters of gross air

309

*Rock shore, Lake Superior Provincial Park*

and water pollution from mining, ore beneficiation, and manufacturing that ran rampant in our immediate past history; there may be clashes, but I think they will be settled in favor of the region's natural values, rather than in favor of corporate profits.

Yet, more insidious threats loom. Chlorinated organic residues—pesticides and industrial chemicals, such as DDT and PCBs—still penetrate the volume of Lake Superior and accumulate in the flesh of our most highly valued fish. Even in Siskiwit Lake on Isle Royale—isolated above the level of Lake Superior since glacial times, and on whose shores there has never been an industrial or agricultural enterprise—the poisons have been detected. Obviously, contamination of the atmosphere is the culprit, with worldwide threats that cannot be brought to bay by North Shore advocates alone.

Acid rain—or acid precipitation, more accurately—continues to produce a snowpack along the shore with the low pH of 3.5, a hundred times more acidic than uncontaminated rainwater. While solutions elude us for now, these problems at least are recognized, by concerned citizens and scientists, if not wholly by governments. I am deeply concerned—but not pessimistic—about these industrial effluents from far away. They are problems that will require the most dedicated efforts from scientists and politicians around the globe, as well as our own vocal support, money, and votes.

However, I am worried most that we may love the North Shore to death. With new recreational technology, demands for easier access and greater comforts, and excessive concentrations of people, the very values of beauty, solitude, and clean environment that are so attractive may themselves be threatened. The problem may be of more immediate concern on the smaller and more congested Minnesota portion, but the potential exists in the Ontario portion, too, if in the more distant future.

But when I see the resolve and talent of my colleagues searching for solutions to problems in aquatic, forest, and wildlife communities; when I see the enthusiasm of park managers in their efforts to preserve wildness along scores of miles of coastline and vast landscapes in hinterlands; when I see fellow anglers, hunters, campers, and others who love special places of wildness, dig into their own pockets to support research programs or donate large segments of their time to organize programs to further public awareness—then I am optimistic and heartened.

It is true that we must provide places where a camper may set up tent or trailer, catch a fish that perhaps started life in a hatchery, and cook it on a charcoal grill. These will continue to be special places for many of us, if only we ensure the integrity of the areas surrounding the campground.

But in our eagerness to protect, we must also take care not to overregiment. There must also be special places—albeit difficult to get to—where a hiker can gather dead wood from the forest, rather than only from the park manager's woodpile. There must be waters where an angler can catch a native brook trout, a hunter shoot a native grouse—and a camper cook this bounty from the land over a campfire.


This final segment is not to be a compendium of all my favorite places, viewing posts, fishing holes, and hiking trails. Rather, it is to suggest that in this vast land north of Superior there is that marvelous opportunity for anyone who wishes to search out and find a special place, or several of them. Not too many areas at all similar to the North Shore offer this opportunity.

Let me suggest, first, a hidden *waterfall*. Of course, the large obvious falls like Kakabeka or Gooseberry are not hidden at all, and many are—like Cross River and Sand River—actually difficult to avoid if you just drive around the shore. But for every falls that is obvious or advertised, perhaps another score occur up- or downstream a ways from the highway or park, by way of a short hike or a secondary road. If you want to be systematic about it, get some topographic maps from the United States Geological Survey or Canadian Department of Energy, Mines and Resources, and calculate the *gradient*, or drop in elevation per length of stream. Wherever the gradient is more than 1 percent—approximately fifty feet drop per mile of stream—a waterfall can be found. Of course, you may find falls in reaches with a gradient lower than that, especially if the drop over your mile is all at one spot!

*Fishing holes.* For brook trout, take along a pocket thermometer on bush-whacking forays; if a seemingly no-account little trickle has summer afternoon temperatures down in the fifties or sixties, and a few spots a foot deep, chances are you'll have some brookies for supper. Or find a beaver pond (fairly new, if possible). On the eastern shore, try small lakes in late summer after everyone else says it is too late for the streams, or anywhere along the shore in reclaimed small lakes, in spring or fall, located with guidemaps or lists available from local fisheries offices. (You will need a canoe.) Good walleye and northern pike lakes are here by the hundreds, but instead try the warmwater rivers, especially in late season "dog days." For rainbows, leave the crowds on the well-known steelhead rivers and explore the small creeks where you will have the one or two pools all to yourself, early in the morning after a warm spring rain or after a fall storm that has broken out the gravel bar at the mouth. For fishing out on the big lake, however, it is best to rely on hired guides and charter captains, at least at first.

*Canoeing.* A number of good lengthy canoe trips, with campsites, are laid out on guides and maps, especially for the eastern shore. At the western end of the shore, the St. Louis and Cloquet are included in Minnesota's system of recreational rivers. All around the shore, many short delightful reaches are available on scores of streams.

Many excellent guidebooks are available for birding, rockhounding and sailing. Search out Indian pictographs, Puckasaw pits, old mine sites. The opportunites for amateur study of geology and botany are endless; start at a provincial or state park with some of the good books available, and work your way out from there. Photography? The North Shore is one of the most photogenic places in the world! Try dawn, foggy days, autumn in the sugar maples, storms, frozen waterfalls.

Seeking solitude? Even on national holidays, with a little effort, you can find some high viewpoint, a bog with sweet gale and orchids, some bit of spooky boreal forest festooned with "old man's beard," a rocky crag with (if you're *very* lucky) a pair of Peregrine Falcons.

When glaciers wasted northward and tundra followed, to be replaced in turn by boreal and pine forests, a few tiny refuges remained behind. These were pockets of northern climate still exposed to harsh conditions, hidden away on islands, windy ledges, shady canyon crevices, or wave-battered shores. Here grow dis-

313

junct plants—those separated from others of their species—surviving in habitats that resemble the Arctic tundra or high mountains. With intriguing names like spikemoss and clubmoss, crowberry, pearlwort and sandwort, and eyebright, they grow untrod yet in the Pukaskwa peninsula, on Neys's rocky coast, in the Susie Islands near Grand Portage, in the depths of Ouimet Canyon. Rare, northern ferns cling to the cool, wet walls of lava gorges.

And a gnarled cedar finds a foothold on a rocky shore.

These special organisms, besides serving as priceless objects of study for ecologists and botanists, present unusual parts of the natural world that interest and excite all of us, and that remind us of our heritage of natural succession on this unique shore—of tundra, forests, and animals—including our own human role in this succession.

Changeable from morning to afternoon, yet unchanging through the centuries, these coastlines and their hinterlands have survived as monuments to a harsh climate and the toughness of ancient rock.

They will command our awe so long as they also command our stewardship.

Then these North Shore strands and beaches will endure as places special to all of us—in remote forests, along unharnessed rivers, and in the wind, waves, and rock upon our wild shores.

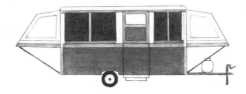

# BIBLIOGRAPHY
# AND
# INDEX

# Bibliography

The Bibliography that follows is organized mainly by *chapter*—that is to say, by *subject*—for the convenience of those who want to read more about a subject discussed in a given chapter. Some chapters dealing with subjects in a common literature—for example, Chapters 3 and 4 on exploration and the fur trade—are combined in one grouping. Because this book is intended to summarize and interpret for general reading, rather than to present the result of scholarly research, the Bibliography mainly contains secondary materials. Included are my main sources and those books and publications more-or-less readily available in bookstores, public libraries, or, for some technical subjects, university libraries.

Several major books cover the geology of the North Shore (Chapter 1). For the Minnesota portion, Duluth to the Pigeon River, the older *Minnesota's rocks and waters,* by George M. Schwartz and George A. Thiel, is still very informative reading, and *Minnesota's geology,* by Richard W. Ojakangas and Charles L. Matsch, presents a newer treatment of northeastern Minnesota, including a stop-by-stop excursion along the North Shore. Ontario's portion is very thoroughly covered by E. G. Pye's excellent guidebook, *Geology and scenery, North Shore of Lake Superior,* from the Pigeon River to Sault Ste. Marie. The geological history of Isle Royale has been meticulously presented by N. King Huber in his *The geologic story of Isle Royale National Park* (see Chapter 16). For a more technical treatment, the symposium edited by R. J. Wold and W. J. Hinze, *Geology and tectonics of the Lake Superior basin,* is the main source for the entire Lake Superior in the early Precambrian Era.

For Chapter 2, a number of books and articles by H. E. Wright, Jr., best document the changing post glacial climate and vegetation in the Lake Superior region; the concurrent development of human cultures in the same area is related by J. V. Wright in his *Ontario prehistory* and by George Irving Quimby in *Indian life in the Upper Great Lakes, 11,000 B.C. to A.D. 1800.*

The eras of exploration and the fur trade, inextricably mixed and so treated together here (Chapters 3 and 4), have been thoroughly researched and reported. A large literature exists. The exhaustive research and delightful writing of Grace Lee Nute stand out; her numerous books on these subjects are still readily available. Marjorie Wilkins Campbell's books on the North West Company and Douglas MacKay's on the Hudson's Bay Company make exciting reading.

The literature on mining in the North Shore region (Chapter 5) is more scattered,

BIBLIOGRAPHY

but two recent books cover well the Mesabi iron ranges: David A. Walker's *Iron frontier* and Edward W. Davis's *Pioneering with taconite*. In a more popular style are Paul deKruif's *Seven iron men* and Stewart H. Holbrook's *Iron brew*. The story of Silver Islet is excitingly portrayed by Helen Moore Strickland in *Silver under the sea*.

The logging history of the North Shore is in two parts—lumber and pulpwood—as are the forests around the shore: the mixed deciduous-coniferous forest containing white and red pine and the boreal forest of spruces (Chapters 6 and 7). For the former, Agnes M. Larson's *History of the white pine industry in Minnesota* is the main source; Jamie Swift's *Cut and run, the assault on Canada's forests* is a sometimes angry account of the pulpwood cutting on the Ontario portion. Stewart H. Holbrook's *Holy old mackinaw*, Donald MacKay's *The lumberjacks*, and J. C. "Buzz" Ryan's several volumes on Minnesota's early loggers give us the color of the men and their tools in the old logging camps. And Clifford and Isabel Ahlgren's *Lob trees in the wilderness* presents an enlightening view of modern forest ecology. Most of the history of the spruce pulpwood harvest on the northern and eastern shores is presented in Canadian government documents.

Published accounts of Lake Superior's fish populations in connection with the sea lamprey invasion (Chapters 8, 9, and 10) are almost entirely from the technical literature. The most complete summaries are in the scientific articles by W. J. Christie, A. H. Lawrie and Jerold F. Rahrer, and Stanford H. Smith. Three voluminous symposiums containing many relevant articles are published as special issues of *Canadian Journal of Fisheries and Aquatic Sciences:* volume 29, number 6 (under former name, *Journal of the Fisheries Research Board of Canada*) on salmonids in lakes; volume 37, number 11 on the sea lamprey and its effects; and volume 38, number 12 on fish stocks, largely in the Great Lakes.

On commercial and sport fishing (Chapters 11 and 12) less is available in the technical literature and more in semi-popular and personal histories. Ed Landin's *A Great Lakes fisherman* stands out for its depiction of the fishermen and their methods of fishing Lake Superior for a living. Historic accounts of commercial fishing are in articles by June Drenning Holmquist, "Commercial fishing on Lake Superior in the 1890s" and Matti Kaups, "North Shore commercial fishing, 1849-1870." Many scientific articles are available on the fishes involved in both commercial and sport fishing, but here I relied heavily on private accounts, written and oral, for the more personal aspects.

For information on birds along the North Shore (Chapter 13) I relied most on *Minnesota birds—where, when, and how many*, by Janet C. Green and Robert B. Janssen, and a number of books on specific or groups of birds: *Gulls: An ecological history*, by Frank Graham, Jr.; *Grouse of the North Shore*, by Gordon Gullion; *Fool hen: The spruce grouse on the Yellow Dog Plains*, by William L. Robinson. Many life history accounts are in the ornithological literature.

Literature on caribou, moose, and white-tailed deer (Chapter 14) is scattered among scientific journals, government publications, and taxonomic books on mammals in general. A notable exception is Randolph L. Peterson's comprehensive book, *North American moose*. Taxonomic books for the mammals include A. W. F. Banfield's *The mammals of Canada*, Evan B. Hazard's *The mammals of Minnesota*, and Randolph L. Peterson's *The mammals of eastern Canada*.

Very little is readily accessible on Lake Nipigon and the Nipigon River (Chapter 15). Here I relied almost entirely on government documents and the scientific literature. Those interested in further reading will do best by inquiring from the Ontario Ministry of Natural Resources, Nipigon, Ontario.

In contrast, Isle Royale (Chapter 16) is the subject of many books and public pamphlets, readily available. The Isle Royale Natural History Association, Houghton, Michigan, has prepared a number of excellent booklets on various elements of the island's plants and animals and human history, such as Peter A. Jordan's on wildlife, Karl F. Lagler and Charles R. Goldman's on fishes, Robert M. Linn's on forests, Peter Oikarinen's on the island's people, Lawrence Rakestraw's two booklets on mining and commercial fishing. Napier Shelton's *The life of Isle Royale* thoroughly covers the general ecology of the island. The moose-wolf story—up to 1979—is comprehensively covered in Durward L. Allen's very readable *Wolves of Minong*. Ingeborg Holte's *Ingeborg's Isle Royale* is a fascinating personal account.

Virtually the entire literature available for parks around the North Shore (Chapter 17) is in government documents and popular magazines. Management plans for the large parks on the Ontario portion are rich with natural resource and historical information, but not very accessible. On the other hand, guidebooks and brochures on specific parks are readily available for the asking, from agencies in both Ontario and Minnesota; most of these are well researched and accurate. See also the list given in the Guides and Maps section of this Bibliography.

Additional references are listed in the General section on some subjects not covered in this book. After the French and English explorations, many British colonials and American travelers explored the wilderness of the northern Great Lakes region and left journals and maps; some of the more notable of these include Jonathan Carver's journals (edited by John Parker); David Thompson's narratives (edited by J. B. Tyrrell); journals and letters of Sir Alexander Mackenzie (edited by W. Kaye Lamb); travel accounts of Commander H. W. Bayfield of the Royal Navy, and Daniel W. Harmon; two volumes by Dr. John Bigsby, *The shoe and canoe;* and the account of the travels of Major Joseph Delafield, *The unfortified boundary,* edited by Robert McElroy and Thomas Riggs.

The story of railroad building on the North Shore region is superbly told by Pierre Berton for the Canadian Pacific Railway and by O. S. Nock for the Algoma Central Railway. (See General section.) Accounts of shipping and shipwrecks on Lake Superior and the other Great Lakes have been presented by Dwight Boyer, *Ships and men of the Great Lakes;* Walter Havighurst, *The long ships passing, the story of the Great Lakes;* William Ratigan, *Great Lakes shipwrecks & survivals;* Frederick Stonehouse, *Isle Royale shipwrecks;* and Julius F. Wolff, *The shipwrecks of Lake Superior.*

A number of local accounts of early settlements are available, usually in the towns and villages concerned, for Beaver Bay by Jessie C. Davis; for Grand Marais by M. J. Humphrey, Ade Toftey, Marion Killmer, and Steven J. Wright, and another by Willis H. Raff; for Marathon by Jean Boultbee; for Tofte by Chris Tormondsen; for Two Harbors by Donna L. Smith; and for Wawa by Agnes W. Turcott. (See General section.)

# *General*

Bayfield, Commander H. W. 1829. Outlines of the geology of Lake Superior. *Transactions of the Literary and Historical Society of Quebec* 1: 1-43.

Berton, Pierre. 1974. *The national dream/The last spike.* (Abridged by author.) Toronto: McClelland and Stewart.

Bertrand, J. P. 1959. *Highway of destiny.* New York. Washington, Hollywood: Vantage Press.

Bigsby, John J. 1850. *The shoe and canoe, or pictures of travel in the Canadas.* Two volumes. London: Chapman and Hall.

Blegen, Theodore C. 1963. *Minnesota, a history of the state.* Minneapolis: University of Minnesota Press.

Boultbee, Jean. 1981. *Pic, pulp, and people.* Revised edition by Jesse Embree. Marathon. Ontario: Township of Marathon. (Original edition 1967. Published by the Marathon Area Centennial Committee.)

Boyer, Dwight. 1977. *Ships and men of the Great Lakes.* New York: Dodd, Mead.

# BIBLIOGRAPHY

Burt, Alfred Leroy. 1944. *A short history of Canada for Americans.* Second edition. Minneapolis: University of Minnesota Press.

Carl, Ray, and James R. Stevens. 1971. *Sacred legends of the Sandy Lake Cree.* Toronto: McClelland and Stewart.

Davis, Jessie C. 1968. *Beaver Bay, original North Shore village.* Duluth: St. Louis County Historical Society.

Duke, Donald, and The Western Writers of America. eds. 1978. *Water trails west.* Garden City, N. Y.: Doubleday.

Fritzen, John. 1974. *Historic sites and place names of Minnesota's North Shore.* Duluth: St. Louis County Historical Society.

Green, Janet C., and John C. Green. 1975. *Inventory of natural, scientific, and aesthetic sites, North Shore of Lake Superior, Minnesota.* Arrowhead Regional Development Commission and Coastal Zone Management Work Group.

Harmon, Daniel W. 1905. *A journal of voyages and travels in the interior of North America.* New York: Allerton.

Havighurst, Walter. 1975. *The long ships passing, the story of the Great Lakes.* Revised edition. New York: Macmillan; London: Collier Macmillan.

Humphrey, M. J., Ade Toftey, Marion Killmer, and Steven J. Wright, eds. 1979. *Pioneer faces and places, 1850-1930.* Grand Marais. Minn.: Cook County Historical Society.

Lake Superior's North Shore. 1974. Special issue, *Ontario Naturalist* 14, no. 4.

Lamb, W. Kaye, ed. 1970. *The journals and letters of Sir Alexander Mackenzie.* London: Cambridge University Press.

Lass, William E. 1977. *Minnesota, a history.* New York and London: Norton.

Life science report, Site Region 4E. Northeastern Region. n.d. Terry Noble, consultant, Thunder Bay. (Prepared for Ontario Ministry of Natural Resources. Unpublished report.)

Litteljohn, Bruce (photography), and Wayland Drew (text). 1975. *Superior: The haunted shore.* Toronto: Gage Publishing.

McElroy, Robert, and Thomas Riggs, eds. 1943. *The unfortified boundary. A diary of the first survey of the Canadian Boundary line from St. Regis to the Lake of the Woods by Major Joseph Delafield.* New York: Privately printed.

Mika, Nick, and Helma Mika. 1972. *Railways of Canada, a pictorial history.* Toronto and Montreal: McGraw-Hill Ryerson.

Munawar, M., ed. 1978. Limnology of Lake Superior. Special issue. *Journal of Great Lakes Research* 4, no. 3-4.

Nock, O.S. 1975. *Algoma Central Railway.* London: A. & C. Black.

Nute, Grace Lee. 1944. *Lake Superior.* Indianapolis and New York: Bobbs-Merrill.

Ontario Department of Public Records and Archives. 1959-83. Press releases on historical plaques: Pigeon River-Grand Portage: Thunder Bay-William McGillivray; Pic Fur post: North West Company post. Sault; Michipicoten Canoe Route; Thunder Bay—Union of North West and Hudson's Bay companies; Savanne Portage; Fort Kaministiquia; Kakabeka Falls.

Parker, John., ed. 1976. *The journals of Jonathan Carver and related documents, 1766-1770.* Bicentennial edition. St. Paul: Minnesota Historical Society Press.

Proceedings of the sea lamprey international symposium (SLIS). 1980. Special issue, *Canadian Journal of Fisheries and Aquatic Sciences* 37, no. 11.

Raff, Willis H. 1981. *Pioneers in the wilderness. Minnesota's Cook County, Grand Marais, and the Gunflint in the 19th century.* Grand Marais. Minn.: Cook County Historical Society.

Ratigan, William. 1960. *Great Lakes shipwrecks & survivals.* New York: Galahad Books.

Schull, Joseph. 1978. *Ontario since 1867.* Toronto: McClelland Stewart.

Searle, R. Newell, and Mark E. Heitlinger. n.d. *Prairies, woods, and islands*. Minneapolis: The Nature Conservancy, Minnesota Chapter.

Smith, Donna L. 1983. *Two Harbors, 100 years*. Two Harbors, Minn.: Lake County Historical Society, and Two Harbors Centennial Commission.

Steinhacker, Charles (photography), and Arno Karlan (text). n.d. (1970?). *Superior*. New York, Evanston, and London: Harper & Row.

Stock concept international symposium (STOCS). 1981. Special issue. *Canadian Journal of Fisheries and Aquatic Sciences* 38, no. 12.

Stonehouse, Frederick. 1983. *Isle Royale shipwrecks*. AuTrain, Mich.: Avery Color Studios.

Swain, Wayland R. 1981. Lake Superior/Lake Baikal. *Lake Superior Port Cities* 3, issue 4: 38-41.

Symposium on salmonid communities in oligotrophic lakes (SCOL). 1972. Special issue, *Journal of the Fisheries Research Board of Canada* 29, no. 6.

Tormondsen, Chris. 1968. *Tofte, a collection of facts and tales of the North Shore area of Lake Superior*. Minneapolis: Hayward-Court Brief Printing.

Turcott, Agnes W. 1982. *Land of the big goose. A history of Wawa and the Michipicoten area, from 1622 to 1982*. Revised edition. Wawa: published by the author.

Tyrrell, J. B., ed. 1916. *David Thompson's narrative of his explorations in western America*. Toronto: The Champlain Society.

Wolff, Dr. Julius F. 1979. *The shipwrecks of Lake Superior*. Duluth: Lake Superior Marine Museum Association.

# Serials

*Canadian Geographic* (formerly *Canadian Geographical Journal*)

*Canadian Journal of Fisheries and Aquatic Sciences* (formerly *Journal of the Fisheries Research Board of Canada*)

*Lake Superior Port Cities*

*The Loon* (formerly *The Flicker*)

*Minnesota History*

*Minnesota Volunteer* (formerly *The Conservation Volunteer*)

*Ontario History*

*Transactions of the American Fisheries Society*

# Guides and Maps

Aguar, Charles E. 1971. *Exploring St. Louis County historical sites*. Duluth: St. Louis County Historical Society.

Baxter, Thomas S. H. 1983. *The traveller's pocket guide to Lake Superior*. Wawa: Superior Lore.

Blakely, Roger. 1981. *Tourist and hiking guide for Minnesota's North Shore*. St. Paul: New Rivers Press.

Bogue, Margaret Beattie, and Virginia A. Palmer. 1979. *Around the shores of Lake Superior: A guide to historic sites*. Madison: University of Wisconsin, Sea Grant College Program.

Breining, Greg, and Linda Watson. 1977. *A gathering of waters, a guide to Minnesota's rivers*. St. Paul: Minnesota Department of Natural Resources.

Buchanan, James W. 1974. *The Minnesota walk book. Vol. 1, A guide to hiking and backpacking in the Arrowhead and Isle Royale*. Duluth: Sweetwater Press.

# BIBLIOGRAPHY

*Canoe routes of Ontario*. 1981. Toronto: Ontario Ministry of Natural Resources, Parks, and Recreational Branch, with McClelland and Stewart.

Dahl, Bonnie. 1983. *The Superior way: A cruising guide to Lake Superior.* Ashland, Wis.: Inland Sea Press.

Daniel, Glenda, and Jerry Sullivan. 1981. *The North Woods of Michigan, Wisconsin, and Minnesota*. San Francisco: Sierra Club Books.

Dorweiler, Paul L. 1982. *Great Lakes trout and salmon fishing*. Coon Rapids, Minn.: Paul Lawrence & Associates.

Duluth Convention and Visitors Bureau and Superior Chamber of Commerce. (1985?). *Twin Ports fishing guide.*

Eckert, Kim R. 1983. *A birder's guide to Minnesota*. Revised second edition. Minneapolis: Minnesota Ornithologists' Union, Bell Museum of Natural History.

*Geological Highway Map, Northern Ontario*. 1980. Ontario Geological Survey, Map 2440. Toronto.

Green, J. C. 1972. *Field trip guidebook for Precambrian North Shore volcanic group, northeastern Minnesota*. Minnesota Geological Survey, University of Minnesota, Guidebook Series no. 3. Minneapolis.

Hereid, Nancy, and Eugene D. Gennaro. 1983. *A family guide to Minnesota's North Shore.* Duluth: University of Minnesota, Sea Grant Extension Program.

Holmquist, June Drenning, and Jean A. Brookins. 1972. *Minnesota's major historic sites, a guide*. Second edition. St. Paul: Minnesota Historical Society.

Lydecker, Ryck. 1976. *The edge of the Arrowhead*. Duluth: Minnesota Marine Advisory Service, Office of Sea Grant.

Minnesota Department of Natural Resources. 1982. *North Shore fishing guide*. St. Paul. (Map, with text.)

Minnesota Department of Natural Resources. 1985. *A guide to lakes managed for stream trout.* St. Paul.

Minnesota Department of Natural Resources. (1985). *North Shore State Trail*. St. Paul. (Map.)

Minnesota Department of Natural Resources. n.d. *Fishing guide to the lakes and streams of Lake County*. St. Paul.

Olsenius, Richard. 1982. *Minnesota travel companion, a guide to the history along Minnesota's highways*. Wayzata, Minn.: Bluestem Productions.

Ontario Ministry of Natural Resources. 1976. *Trout and salmon migratory routes, northern Ontario streams*. Toronto. (Map.)

Ontario Ministry of Natural Resources. 1978-79. *Winter outdoors: Ontario provincial parks and trails*. Toronto.

Pye, E. G. 1968. *Geology and scenery, Rainy Lake and east to Lake Superior.* Ontario Department of Mines, Geological Guide Book no 1. Toronto.

Pye, E. G. (1969) 1975. *Geology and scenery, North Shore of Lake Superior.* Reprint. Ontario Department of Mines, Geological Guide Book no 2. Toronto.

Sansome, Constance J. 1983. *Minnesota underfoot, a field guide to the state's outstanding geologic features*. Bloomington, Minn.: Voyageur Press.

Umhoefer, Jim. 1984. *Guide to Minnesota's parks, canoe routes, and trails*. Madison, Wis.: Northword.

# *Chapter 1. Genesis by Fire*

Agassiz, Louis. 1850. *Lake Superior: Its physical character, vegetation, and animals, compared with those of other and similar regions*. With a narrative of the tour, by J. Elliot Cabot. (Reprinted 1974, New York: Robert E. Krieger Publishing, Facsimile.) Boston: Gould, Kendall and Lincoln.

Billings, Michael Dennis. 1974. A survey of the geology and geomorphology of the Slate Islands, Lake Superior, District of Thunder Bay, Ontario. Toronto:Ontario Ministry of Natural Resources, Park Planning Branch. (Unpublished report.)

Bostock, H. S. 1970. Physiographic subdivisions of Canada. In *Geology and economic minerals of Canada*, fifth edition, ed. R. J. W. Douglas, pp. 9-30. Geological Survey of Canada, Economic Geology Report no. 1.

Bray, Edmund C. 1962. *A million years in Minnesota, the glacial story of the state.* St. Paul: Science Museum of Minnesota.

Bray, Edmund C. 1977. *Billions of years in Minnesota, the geological story of the state.* St. Paul: Science Museum of Minnesota.

Clayton, Lee. 1983. Chronology of Lake Agassiz drainage to Lake Superior. In *Glacial Lake Agassiz*, ed. James T. Teller and Lee Clayton, pp. 291-307. Geological Association of Canada, Special Paper 26. Boulder, Colo.

Coleman, A. P. 1922. *Glacial and pre-glacial lakes in Ontario.* University of Toronto Studies: Publications of the Ontario Fisheries Research Laboratory, no. 10.

Douglas, R. J. W., ed. 1970. *Geology and economic minerals of Canada.* Geological Survey of Canada, Economic Geology Report no. 1. Ottawa.

Flint, Richard Foster. 1947. *Glacial geology and the Pleistocene Epoch.* New York: John Wiley.

Green, J. C. 1972. General geology, Northeastern Minnesota. In *Geology of Minnesota: A centennial volume*, ed. P. K. Sims and G. B. Morey, pp. 292-93. St. Paul: Minnesota Geological Survey.

Green, J. C. 1972. North Shore volcanic group. In *Geology of Minnesota: A centennial volume*, ed. P. K. Sims and G. B. Morey, pp. 294-332. St. Paul: Minnesota Geological Survey.

Green, J. C. 1978. Why is Lake Superior? *Minnesota Volunteer* 41, no. 239 (July-August): 10-19.

Green, J. C., Mark A. Jirsa, and Carol M. Moss. 1977. *Environmental geology of the North Shore.* St. Paul: Minnesota Geological Survey.

Hewitt, D. F., and E. B. Freeman. 1972. *Rocks and minerals of Ontario.* Ontario Department of Mines and Northern Affairs, Geological Circular 13. Toronto.

Hough, J. L. 1958. *Geology of the Great Lakes.* Urbana: University of Illinois Press.

Klasner, J. S., W. F. Cannon, and W. R. Van Schmus. 1982. The pre-Keweenawan tectonic history of southern Canadian Shield and its influence on formation of the Midcontinent Rift. In *Geology and tectonics of the Lake Superior basin*, ed. R. J. Wold and W. J. Hinze, pp. 27-46. Geological Society of America Memoir 156. Boulder, Colo.

Morey, G. B. 1972. Mesabi Range. In *Geology of Minnesot: A centennial volume*, ed. P. K. Sims and G. B. Morey, pp. 204-17. St. Paul: Minnesota Geological Survey.

Ojakangas, Richard W., and Charles L. Matsch. 1982. *Minnesota's geology.* Minneapolis: University of Minnesota Press.

Phillips, Brian A. M. 1975. How molten rocks and lava and then ice sheets shaped the grandeur of Superior's north shore. *Canadian Geographical Journal* 91, no. 5 (November): 4-11, 49.

Prest, V. K. 1970. Quaternary geology of Canada. In *Geology and economic minerals of Canada*, fifth edition, ed. R. J. W. Douglas, pp. 675-764. Geological Survey of Canada, Economic Geology Report no. 1.

Pye, E. G. (1969) 1975. *Geology and scenery, North Shore of Lake Superior.* Reprint. Ontario Department of Mines, Geological Guide Book no. 2. Toronto.

Saarnisto, Matti. 1974. The deglaciation history of the Lake Superior region and its climatic implications. *Quaternary Research* 4: 316-39.

BIBLIOGRAPHY

Schwartz, George M., and George A. Thiel. 1963. *Minnesota's rocks and waters, a geological story*. Revised edition. Minneapolis: University of Minnesota Press.

Sims, P. K., and G. B. Morey, eds. 1972. *Geology of Minnesota: A centennial volume*. St. Paul: Minnesota Geological Survey.

Stockwell, C. H., J. C. McGlynn, R. F. Emslie, B. V. Sanford, A. W. Norris, J. A. Donaldson, W. F. Fahrig, and K. L. Currie. 1970. Geology of the Canadian Shield. In *Geology and economic minerals of Canada*, ed. R. J. W. Douglas, pp. 43-150. Geological Survey of Canada, Economic Geology Report no. 1.

Thiel, George A. 1947. *The geology and underground waters of northeastern Minnesota*. Minneapolis: University of Minnesota Press.

Van Hise, C. R., and C. K. Leith. 1911. *The geology of the Lake Superior region*. U. S. Geological Survey Monograph 52. Washington, D.C.

Wold, R. J., and W. J. Hinze, eds. 1982. *Geology and tectonics of the Lake Superior basin*. Geological Society of America Memoir 156. Boulder, Colo.

Wright, H. E., Jr. 1976. Ice retreat and revegetation in the western Great Lakes area. In *Quaternary stratigraphy of North America*, ed. W. C. Mahaney, pp. 119-32, Stroudsburg, Penn.: Dowden, Hutchinson & Ross.

Wright, H. E., Jr., ed. 1983. *Late-Quaternary environments of the United States*. Vol. 1, *The late Pleistocene*, ed. Stephen C. Porter. Vol. 2, *The Holocene*, ed. H. E. Wright, Jr. Minneapolis: University of Minnesota Press.

Wright, H. E., Jr., and William A. Watts. 1969. *Glacial and vegetational history of northeastern Minnesota*. University of Minnesota, Minnesota Geological Survey, Special Publication Series SP-11. Minneapolis.

Zaslow, Morris. 1975. *Reading the rocks*. Toronto: Macmillan; Ottawa: Department of Energy, Mines, and Resources.

# Chapter 2. The Early Americans

Allin, Albert E. 1959. The North Shore's changing wildlife. *Naturalist* 10(4): 24-29.

Bishop, Charles A. 1974. *The northern Ojibwa and the fur trade: An historical and ecological study*. Toronto and Montreal: Holt, Rinehart and Winston.

Black, Linda. 1980. The prehistory of the north central region of Ontario. Toronto. Ontario Ministry of Natural Resources. (Unpublished report.)

Canby, Thomas Y. 1979. The search for the first Americans. *National Geographic* 156: 330-63.

Cleland, Charles Edward. 1966. *The prehistoric animal ecology and ethnozoology of the Upper Great Lakes region*. Museum of Anthropology, University of Michigan, Anthropological Papers no. 29. Ann Arbor.

Cleland, Charles Edward. 1983. Indians in a changing environment. In *The Great Lakes forest: An environmental and social history*, ed. Susan L. Flader, pp. 83-95. Minneapolis: University of Minnesota Press.

Coatsworth, Emerson S. 1970. *Nomads of the Shield: Ojibwa Indians*. Canada: Ginn.

Conway, Thor. 1975. An archaeological survey of the northeastern shore of Lake Superior. Toronto: Ontario Ministry of Natural Resources. (Unpublished report.)

Conway, Thor. 1981. *Archaeology in northeastern Ontario, searching for our past*. Ottawa: Ontario Ministry of Culture and Recreation.

Cushing, E. H., and H. E. Wright, Jr., eds. 1967. *Quaternary paleoecology*. New Haven: Yale University Press.

Danziger, Edmund Jefferson, Jr. 1979. *The Chippewas of Lake Superior*. Norman: University of Oklahoma Press.

# BIBLIOGRAPHY

Deevey, Edward S., Jr. 1949. Biogeography of the Pleistocene. *Bulletin of the Geological Society of America* 60: 1315-1416.

Fries, Magnus. 1962. Pollen profiles of late Pleistocene and recent sediments from Weber Lake, Minnesota. *Ecology* 43: 295-308.

Johnson, Elden. 1978. *The prehistoric peoples of Minnesota*. Second edition. Minnesota Prehistoric Archaeology Series no. 3. St. Paul: Minnesota Historical Society.

Marie-Victorin, Frere. 1938. Phytogeographical problems of eastern Canada. *American Midland Naturalist* 19: 489-558.

Marsh, John S. 1976. The human history of the Pukaskwa National Park area, 1650 to 1975. Ottawa: Parks Canada, Pukaskwa National Park. (Unpublished manuscript.)

O'Sullivan, Thomas. 1982. Artists portray North Shore's Witch Tree. *Minnesota Volunteer* 45, no. 263 (July-August): 48-51.

Peters, Gordon R. 1981. Tracking Minnesota's ancient people. *Minnesota Volunteer* 44, no. 257 (July-August): 48-53.

Quimby, George Irving. 1960. *Indian life in the Upper Great Lakes, 11,000 B.C. to A.D. 1800.* Chicago: University of Chicago Press.

Ritzenthaler, Robert E., and Pat Ritzenthaler. 1970. *The Woodland Indians of the Western Great Lakes.* Garden City, N. Y.: Natural History Press.

Rostlund, Erhard. 1952. *Freshwater fish and fishing in native North America*. Berkeley and Los Angeles: University of California Press.

Terasmae, J. 1967. Postglacial chronology and forest history in the northern Lake Huron and Lake Superior regions. In *Quaternary paleoecology*, ed. E. H. Cushing and H. E. Wright, Jr., pp. 45-58. New Haven: Yale University Press.

Webers, Gerald F., and Bruce R. Erickson. 1984. Minnesota's Ice Age mammals. *Minnesota Volunteer* 47, no. 275 (July-August): 55-61.

Wright, H. E., Jr. 1964. Aspects of early postglacial forest succession in the Great Lakes region. *Ecology* 45: 439-48.

Wright, H. E., Jr. 1968. The roles of pine and spruce in the forest history of Minnesota and adjacent areas. *Ecology* 49: 937-55.

Wright, H. E., Jr. 1976. Ice retreat and revegetation in the western Great Lakes area. In *Quaternary stratigraphy of North America,* ed. W. C. Mahaney, pp. 119-32. Stroudsburg, Penn.: Dowden, Hutchinson & Ross.

Wright, H. E., Jr., ed. 1983. *Late-Quaternary environments of the United States*. Vol. 1, *The late Pleistocene,* ed. Stephen C. Porter. Vol. 2, *The Holocene,* ed. H. E. Wright, Jr. Minneapolis: University of Minnesota Press.

Wright, H. E., Jr., and William A. Watts. 1969. *Glacial and vegetational history of northeastern Minnesota*. University of Minnesota, Minnesota Geological Survey Special Publication no. 11. Minneapolis.

Wright, J. V. 1963. *An archaeological survey along the North Shore of Lake Superior.* National Museum of Canada, Anthropology Paper no. 3. Ottawa.

Wright, J. V. 1972. *Ontario prehistory, an eleven-thousand-year archaeological outline*. Ottawa: National Museum of Man, National Museums of Canada, Archaeological Survey of Canada.

Wright, J. V. 1976. *Six chapters of Canada's prehistory*. Ottawa: National Museums of Canada, National Museum of Man, Archaeological Survey of Canada.

# BIBLIOGRAPHY

# Chapter 3. The Northwest Adventurers
# Chapter 4. Empires of Fur Posts

Bain, James, ed. 1901. *Travels and adventures in Canada and the Indian territories, between the years 1760 and 1776, by Alexander Henry, fur trader.* Boston: Little, Brown.

Baldwin, Doug. 1974. *The fur trade in the Moose-Missinaibi River Valley, 1770-1917.* Toronto: Ontario Ministry of Natural Resources, Division of Parks, Historical Sites Branch.

Bishop, Charles A. 1974. *The northern Ojibwa and the fur trade: An historical and ecological study.* Toronto, Montreal: Holt, Rinehart and Winston.

Bissell, Orpha Lucetta. 1917. The activities of the American Fur Company. M.A. thesis, University of Minnesota. Minneapolis.

Bolz, J. Arnold. 1960. *Portage into the past.* Minneapolis: University of Minnesota Press.

Boultbee, Jean. 1981. *Pic, pulp, and people.* Revised edition by Jesse Embree. Marathon, Ontario: Township of Marathon. (Original edition 1967. Published by the Marathon Area Centennial Committee.)

Brown, Jennifer S. H. 1980. *Strangers in blood; fur trade company families in Indian country.* Vancouver and London: University of British Columbia Press.

Browne, George Waldo. 1905. *The St. Lawrence River.* New York: Weathervane Books.

Buck, Solon J. 1931. *The story of the Grand Portage.* Minneapolis: Cook County Historical Society.

Buckley, Thomas C., ed. 1984. *Rendezvous, selected papers of the Fourth North American Fur Trade Conference, 1981.* St. Paul: North American Fur Trade Conference.

Burpee, Lawrence J. (1933) 1963. La Vérendrye—pathfinder of the west. Reprint. *Canadian Geographical Journal* 66, no. 1 (January): 44-49.

Campbell, George. 1984. Voyageurs will rendezvous at (New) Old Fort William. *Lake Superior Port Cities* 5, issue 4 (Spring): 17-20, 68-69.

Campbell, Marjorie Wilkins. 1957. *The North West Company.* Toronto: Macmillan.

Campbell, Marjorie Wilkins. 1974. *The Nor'Westers, the fight for the fur trade.* Toronto: Macmillan.

Campbell, Susan. 1980. *Fort William, Living and working at the post.* Thunder Bay: Ontario Ministry of Culture and Recreation.

Dawson, K. C. A. 1969. *A report on Lake Nipigon fur trading posts.* Toronto: Ontario Archaeological and Historic Sites Board.

Eddington, Bryan. 1981. The Great Hall at Old Fort William. *The Beaver.* Summer: 38-42.

Fritzen, John. 1978. *The history of Fond du Lac and Jay Cooke Park.* Duluth: St. Louis County Historical Society.

Gall, Patricia L. 1967. The excavation of Fort Pic, Ontario. *Ontario Archeology* 10: 34-63.

Gates, Charles M., ed. (1933) 1965. *Five fur traders of the Northwest.* Reprint. St. Paul: Minnesota Historical Society.

Gilman, Carolyn. 1982. *Where two worlds meet, the Great Lakes fur trade.* St. Paul: Minnesota Historical Society.

Griffith, Ray. 1972. The history of Lake Nipigon. Toronto: Ontario Ministry of Natural Resources. (Unpublished report.)

Heidenreich, Conrad E. 1980. Mapping the Great Lakes/The period of exploration, 1603-1700. *Cartographica* 17, no. 3: 32-64.

Hind, Henry Youle. 1860. *Narrative of the Canadian Red River exploring expedition of 1857 and of the Assiniboine and Saskatchewan exploring expedition of 1858.* Two volumes. London: Longman, Green, Longman, and Roberts.

Huot-Vickery, Jim Dale. 1982. Retracing the Kam-Dog in northwestern Ontario. *Canadian Geographic* 102(4): 32-37.

Innis, Harold A. 1956. *The fur trade in Canada*. Revised edition. Toronto and Buffalo: University of Toronto Press.

Kellogg, Louise Phelps, ed. 1917. *Early narratives of the Northwest, 1634-1699*. New York: Charles Scribner's Sons.

Lass, William E. 1980. *Minnesota's boundary with Canada, its evolution since 1783*. St. Paul: Minnesota Historical Society Press.

Lavender, David. 1975. *The fist in the wilderness*. Albuquerque: University of New Mexico Press.

Lindquist, Maude L. 1976. The Lake Superior country; magnet for power. In *Duluth, sketches of the past, a bicentennial collection*, ed. Ryck Lydecker and Lawrence J. Sommer, pp. 8-16. Duluth: American Revolution Bicentennial Commission.

MacDonald, Graham A. 1974. *East of Superior: A history of the Lake Superior Provincial Park region*. Wawa: Ontario Ministry of Natural Resources, Parks Branch.

Macgillivray, George B. 1970. *Our heritage, a brief history of early Fort William and the great North West Company, its personalities and competitors, from 1764 to 1830*. Dryden, Ontario: Alex Wilson Publications.

MacKay, Douglas. 1966. *The honourable company*. Revised edition. Toronto and Montreal: McClelland and Stewart.

MacLennan, Hugh. 1961. *Seven rivers of Canada*. Toronto: Macmillan of Canada.

Martin, Calvin. 1978. *Keepers of the game, Indian-animal relationships and the fur trade*. Berkeley: University of California Press.

McCaig, Robert. 1978. The great canoe trail. In *Water trails west*, ed. Donald Duke and The Western Writers of America, pp. 1-14. Garden City, N.Y.: Doubleday.

McComber, A. J. 1923/1924. Some early history of Thunder Bay and District. *Thunder Bay Historical Society, Annual Reports and Papers* 15: 13-22.

McElroy, Robert, and Thomas Riggs, eds. 1943. *The unfortified boundary. A diary of the first survey of the Canadian Boundary line from St. Regis to the Lake of the Woods by Major Joseph Delafield*. New York: privately printed.

Morse, Eric W. 1962. *Canoe routes of the voyageurs, the geography and logistics of the Canadian fur trade*. Toronto: Quetico Foundation of Ontario; St. Paul: Minnesota Historical Society.

Morse, Eric W. 1979. *Fur trade canoe routes of Canada/then and now*. Toronto, Buffalo, London: University of Toronto Press.

Morton, W. L. 1966. The North West Company, pedlars extraordinary. *Minnesota History* 40: 157-65.

Nute, Grace Lee. 1930. Posts in the Minnesota fur-trading area, 1660-1855. *Minnesota History* 11:353-85.

Nute, Grace Lee. 1941. *The voyageur's highway: Minnesota's border lake land*. St. Paul: Minnesota Historical Society.

Nute, Grace Lee. 1944. *Lake Superior*. Indianapolis and New York: Bobbs-Merrill.

Nute, Grace Lee. 1950. *Rainy River Country: A brief history of the region bordering Minnesota and Ontario*. St. Paul: Minnesota Historical Society.

Nute, Grace Lee. 1955. *The voyageur*. Reprint edition. St. Paul: Minnesota Historical Society.

Nute, Grace Lee. 1959. Three centuries ago. *Naturalist* 10, issue 4 (Winter): 3-6.

Nute, Grace Lee. 1978. *Caesars of the wilderness*. Reprint edition. St. Paul: Minnesota Historical Society Press.

Ontario Department of Lands and Forests. 1965. *A history of Sault Ste. Marie Forest District*. District History Series, no. 20. Toronto.

Ontario Ministry of Natural Resources. n.d. *The boundary waters fur trade canoe route.* Toronto.

Ontario Ministry of Natural Resources. n.d. *The Kaministikwia River fur trade canoe route.* Toronto.

Rich, E. E. 1967. *The fur trade and the Northwest to 1857.* Toronto: McClelland and Stewart.

Roberts, Leslie. 1949. *The Mackenzie.* New York: Rinehart.

Saunders, R. M. 1940. Coureur des bois: A definition. *The Canadian Historical Review* 21: 123-31.

Sommer, Lawrence J., ed. 1979. *Daniel Greysolon, Sieur duLhut, A tercentenary tribute.* Duluth: St. Louis County Historical Society.

Thwaites, Reuben Gold, ed. 1896-1901. *The Jesuit Relations and allied documents. Travels and explorations of the Jesuit Missionaries in New France, 1610-1791.* 73 volumes. Cleveland: Burrows Brothers.

Treuer, Robert. 1979. *Voyageur country.* Minneapolis: University of Minnesota Press.

Wallace, W. Stewart. 1954. *The pedlars from Quebec, and other papers on the Nor'Westers.* Toronto: Ryerson Press.

Weiler, John. 1974. Michipicoten, a historic landmark. *Ontario Naturalist,* December: 11-13.

Weiler, John. 1980. Michipicoten: Hudson's Bay Company post, 1821-1904. In *Three heritage studies, on the history of the HBC Michipicoten post and on the archaeology of the North Pickering area,* ed. David Skene Melvin, pp. 4-64. Ontario Ministry of Culture and Recreation, Historical Planning and Research Branch, Archaeological Research Report 14, Heritage Planning Study 5. Toronto.

Wheeler, Robert C., Walter A. Kenyon, Alan R. Woolworth, and Douglas A. Birk. 1975. *Voices from the rapids, an underwater search for fur trade artifacts, 1960-73.* St. Paul: Minnesota Historical Society.

Williams, Glyndwr. 1983. The Hudson's Bay Company and the fur trade: 1670-1870. Special issue, *The Beaver,* Autumn.

Wilson, Clifford P. 1963. Étienne Brulé and the Great Lakes. *Canadian Geographical Journal* 66, no. 1 (January): 38-43.

Woolworth, Alan R. 1982. The great carrying place, Grand Portage. In *Where two worlds meet, the Great Lakes fur trade,* ed. Carolyn Gilman, pp. 110-15. St. Paul: Minnesota Historical Society.

Woolworth, Nancy L. 1975. Grand Portage in the Revolutionary War. *Minnesota History* 44: 199-208.

# Chapter 5. Iron Mountains and Other Lodes

Bartlett, Robert V. 1980. *The Reserve Mining controversy: Science, technology, and environmental quality.* Bloomington: Indiana University Press.

Brown, Eldon L., and W. F. Morrison. 1942. Geology of the Josephine mine. *Canadian Mining Journal* 63: 5-9.

Collins, W. H., T. T. Quirke, and Ellis Thompson. 1926. *Michipicoten iron ranges.* Canada Department of Mines, Geological Survey, Memoir 147. Ottawa.

Davis, Edward W. 1953. Taconite, the derivation of the name. *Minnesota History* 33: 282-83.

Davis, Edward W. 1964. *Pioneering with taconite.* St. Paul: Minnesota Historical Society.

deKruif, Paul. 1929. *Seven iron men.* New York: Harcourt, Brace.

Dorin, Patrick C. 1977. *The Lake Superior iron ore railroads.* Second edition. New York: Bonanza Books, Crown Publishers.

Eliseuson, Michael. 1976. *Tower Soudan: The state park down under.* St. Paul: Minnesota Parks Foundation.

Hatcher, Harlan. 1950. *A century of iron and men*. Indianapolis and New York: Bobbs-Merrill.

Hatcher, Harlan, and Erich A. Walter. 1963. *A pictorial history of the Great Lakes*. New York: American Legacy Press.

Hewitt, D. F., and E. B. Freeman. 1972. *Rocks and minerals of Ontario*. Ontario Department of Mines and Northern Affairs, GC 13. Toronto.

Holbrook, Stewart H. 1939. *Iron brew: A century of American ore and steel*. New York: Macmillan.

Kraske, Robert. 1979. "Hey, the gold fields, boys!" *Minnesota Volunteer* 42, no. 247 (November-December): 22-31.

Lipp, Rodney J. 1985. *Minnesota Mining Directory*. Minneapolis: Mineral Resources Research Center, University of Minnesota.

Marshall, James R. 1982. Silver Islet: Quest for Superior's riches. *Lake Superior Port Cities* 4, issue 1: 20-24, 66-67.

Michipicoten, Township of. 1979. *Heritage Michipicoten*.

Morehouse, William D. 1983. Where the bonanza trail began. In *Ontario, the pioneer years*, ed. T. W. Paterson, pp. 78-92. Toronto: Cannon Book Distribution.

Morey, G. B. 1973. Mesabi, Gunflint, and Cuyuna ranges, Minnesota. In *Genesis of Precambrian iron and manganese deposits*, (Proceedings Kiev Symposium, 1970), pp. 193-208. Earth Sciences, 9. UNESCO.

Nock, O. S. 1975. *Algoma Central Railway*. London: A & C Black.

Ontario Ministry of Natural Resources. 1979. *Ontario mining statistics, a preliminary compendium*. Toronto: Mineral Resources Branch and Centre for Resource Studies.

Patterson, G. C. 1983. Exploration history in the Hemlo area. In *The geology of gold in Ontario*, ed. A. C. Colvine, pp. 226-29. Ontario Geological Survey, Miscellaneous Paper 110. Toronto.

Patterson, G. C. 1984. *Field trip guidebook to the Hemlo area*. Ontario Geological Survey, Miscellaneous Paper 118. Toronto.

Skillings, Helen Wieland. 1972. *We're standing on iron!* Duluth: St. Louis County Historical Society.

Strickland, Helen Moore. 1979. *Silver under the sea*. Cobalt, Ontario: Highway Book Shop.

Thompson, Pamela M., ed. 1979. *Iron Range country: A historical travelogue of Minnesota's iron ranges*. Eveleth, Minn.: Iron Range Resources and Rehabilitation Board.

Van Emery, Margaret. 1964. Francis Hector Clergue and the rise of Sault Ste. Marie as an industrial centre. *Ontario History* 56: 191-202.

Walker, David A. 1979. *Iron frontier: The discovery and early development of Minnesota's three ranges*. St. Paul: Minnesota Historical Society.

Wroe, John. 1985-1986. Gold hurry! Extracting bullion worth billions. *Canadian Geographic* 105, no. 6 (December-January): 20-31.

# *Chapter 6. The Tall Pines*
# *Chapter 7. The Boreal Forest*

Ahlgren, Clifford E., and Isabel F. Ahlgren. 1981. Comeback for the white pine? *Minnesota Volunteer* 44, no. 258 (September-October): 37-43.

Ahlgren, Clifford E., and Isabel F. Ahlgren. 1984. *Lob trees in the wilderness*. Minneapolis: University of Minnesota Press.

Barr, Elinor. 1976. Lumbering in the Pigeon River watershed. *Thunder Bay Historical Museum Society, Papers and Records* 4: 3-9.

# BIBLIOGRAPHY

Beggs, Robert C. 1982. A chronological record of the developments in harvesting systems on the Dog River Limits of Great Lakes Forest Products. B.S. thesis, School of Forestry, Lakehead University, Thunder Bay.

Bertrand, J. P. (1960?). Timber wolves. Thunder Bay: Lakehead University Library. (Unpublished manuscript.)

Boultbee, Jean. 1981. *Pic, pulp and people*. Revised edition by Jesse Embree. Marathon, Ontario: Township of Marathon. (Original edition 1967. Published by the Marathon Area Centennial Committee.)

Brown, Nelson Courtland. 1936. *Logging—transportation*. New York: John Wiley.

Buchan, J. D. 1972. *Logging of the Black and White watersheds: The pre-mechanization era, 1890-1950*. Toronto: Ontario Ministry of Natural Resources. (Unpublished report.)

Crossen, T. Wayne. 1976. *Nicholson: A study of lumbering in North Central Ontario, 1885-1930, with special reference to the Austin-Nicholson Company*. Toronto: Ontario Ministry of Culture and Recreation, for Ministry of Natural Resources. (Unpublished report.)

Defebaugh, James Elliott. 1906. *History of the lumber industry of America*. 1. Chicago: American Lumberman.

Dobie, James S. 1937. The past half-century in Northern Ontario. *Proceedings of the Association of Ontario Land Surveyors* 45: 138-56.

Eliseuson, Michael. 1986. The paper world of Lake Superior. *Lake Superior Port Cities* 8, issue 3 (May-June): 21-24.

Fahl, Ronald J. 1977. *North American forest and conservation history: A bibliography*. Santa Barbara, Calif.: A.B.C.—Clio Press.

Flader, Susan L., ed. 1983. *The Great Lakes forest, An environmental and social history*. Minneapolis: University of Minnesota Press.

Fritzen, John. 1968. History of the North Shore lumbering. Duluth: St. Louis County Historical Society. (Unpublished report.)

Gollat, R. 1976. The fish and wildlife values of the Black Bay peninsula and offshore Lake Superior islands. Toronto: Ontario Ministry of Natural Resources. (Unpublished report.)

Hedstrom, Margaret. 1974. *Hedstrom Lumber Company after 60 years, 1914-1974*. Grand Marais, Minn.: Hedstrom Lumber.

Holbrook, Stewart H. 1939. *Holy old Mackinaw*. New York: Macmillan.

Kauffmann, Carl. 1970. *Logging days in Blind River*. Sault Ste. Marie, Ontario: Carl Kauffmann.

King, Frank A. 1981. *Minnesota logging railroads*. San Marino, Calif.: Golden West Books.

Lambert, Richard S. 1967. *Renewing nature's wealth*. Toronto: Ontario Department of Lands and Forests.

Larson, Agnes M. 1949. *History of the white pine industry in Minnesota*. Minneapolis: University of Minnesota Press.

Lower, A. R. M., W. A. Carrothers, and S. A. Saunders. 1938. *The North American assault on the Canadian forest*. Toronto: Ryerson Press.

MacKay, Donald. 1978. *The lumberjacks*. Toronto: McGraw-Hill Ryerson.

Mika, Nick, and Helma Mika. 1972. *Railways of Canada: A pictorial history*. Toronto and Montreal: McGraw-Hill Ryerson.

Nasky, Frank. 1983. *Industry in the wilderness: The people, the buildings, the machines—heritage in Northwestern Ontario*. Toronto and Charlottetown, Ontario: Dundurn Press.

Ontario Department of Lands and Forests. 1965. *A history of Sault Ste. Marie Forest District*. District History Series, no. 20. Toronto.

Ontario Ministry of Natural Resources. 1981a. *Chapleau District, Background Information*. Toronto.

Ontario Ministry of Natural Resources. 1981b. *Nipigon land use plan, background information.* Toronto.

Rector, William Gerald. 1953. *Log transportation in the lake states lumber industry, 1840-1918.* Glendale, Calif.: Arthur H. Clark.

Rowe, J. S. 1959. *Forest regions of Canada.* Canada Department of Northern Affairs and National Resources, Forestry Branch, Bulletin 123. Ottawa.

Ryan, J. C. 1973. *Early loggers in Minnesota.* Duluth: Minnesota Timber Producers.

Ryan, J. C. 1976. *Early loggers in Minnesota—Vol. II.* Duluth: Minnesota Timber Producers.

Ryan, J. C. 1980. *Early loggers in Minnesota—Vol. III.* Duluth: Minnesota Timber Producers.

Swift, Jamie. 1983. *Cut and run, the assault on Canada's forests.* Toronto: Between the Lines.

Vosper, G. F. 1984. *Logging in Lake Superior Provincial Park.* Toronto: Ontario Ministry of Natural Resources, Wawa District.

# Chapter 8. Fish Story I
# Chapter 9. Invader from the Sea
# Chapter 10. Fish Story II

Anderson, Emory J., and Lloyd L. Smith, Jr. 1971. *A synoptic study of food habits of 30 fish species from western Lake Superior.* University of Minnesota Agricultural Experiment Station, Technical Bulletin 279. St. Paul.

Applegate, Vernon C. 1950. *Natural history of the sea lamprey (Petromyzon marinus) in Michigan.* U.S. Fish and Wildlife Service, Special Scientific Report: Fisheries no. 55. Washington, D.C.

Applegate, Vernon C., John H. Howell, A. E. Hall, Jr., and Manning A. Smith. 1957. *Toxicity of 4,346 chemicals to larval lampreys and fishes.* U.S. Fish and Wildlife Service, Special Scientific Report: Fisheries no. 207. Washington, D.C.

Aron, William I., and Stanford H. Smith. 1971. Ship canals and aquatic ecosystems. *Science* 174, no. 4004: 13-20.

Bailey, Reeve M., and Gerald R. Smith. 1981. Origin and geography of the fish fauna of the Laurentian Great Lakes basin. *Canadian Journal of Fisheries and Aquatic Sciences* 38: 1539-61.

Christie, W. J. 1974. Changes in the fish species composition of the Great Lakes. *Journal of the Fisheries Research Board of Canada* 31: 827-54.

Christie, W. J., J. M. Fraser, and S. J. Nepszy. 1972. Effects of species introductions on salmonid communities in oligotrophic lakes. *Journal of the Fisheries Research Board of Canada* 29: 969-73.

Dahl, F. H., and R. B. McDonald. 1980. Effects of control of the sea lamprey (*Petromyzon marinus*) on migratory and resident fish populations. *Canadian Journal of Fisheries and Aquatic Sciences* 37: 1886-94.

Downs, Warren. 1976. *Fish of Lake Superior.* University of Wisconsin Sea Grant Program, WIS-SG-76-124. Madison.

Downs, Warren. 1982. *The sea lamprey: Invader of the Great Lakes.* University of Wisconsin Sea Grant Institute, Great Lakes Alien Series no. 1. Madison.

Eddy, Samuel, and James C. Underhill. 1974. *Northern fishes.* Third edition. Minneapolis: University of Minnesota Press.

Emery, Lee. 1985. *Review of fish species introduced into the Great Lakes, 1819-1974.* Great Lakes Fishery Commission, Technical Report no. 45. Ann Arbor.

# BIBLIOGRAPHY

Goodier, John Lawrence. 1981. Native lake trout *(Salvelinus namaycush)* stocks in the Canadian waters of Lake Superior prior to 1955. M.S. thesis, University of Toronto. (Abridged version in *Canadian Journal of Fisheries and Aquatic Sciences* 38:1724-37.)

Grant, Alexander J. 1930. Welland Ship Canal. In *Welland Ship Canal,* ed. C. M. Pearsall, pp. 5-13. St. Catharines, Ontario: Commercial Press.

Great Lakes Fishery Commission. 1983. *Annual report for the year 1981.* Ann Arbor.

Hassinger, Richard L. 1978. Fishing the Great Lake. *Minnesota Volunteer* 41, no. 239: 20-27.

Hubbs, Carl L., and T. E. B. Pope. 1937. The spread of the sea lamprey through the Great Lakes. *Transactions of the American Fisheries Society* 66: 172-76.

Jackson, John N. 1975. *Welland and the Welland Canal, the Welland Canal by-pass.* Belleville, Ontario: Mika Publishing.

Krueger, Charles C., and George R. Spangler. 1981. Genetic identification of sea lamprey *(Petromyzon marinus)* populations from the Lake Superior basin. *Canadian Journal of Fisheries and Aquatic Sciences* 38: 1832-37.

Kwain, Wen-hwa, and Andrew H. Lawrie. 1981. Pink salmon in the Great Lakes. *Fisheries* 6: 2-6.

Lark, J. G. I. 1973. An early record of the sea lamprey *(Petromyzon marinus)* from Lake Ontario. *Journal of the Fisheries Research Board of Canada* 30: 131-33.

Lawrie, A. H. 1970. The sea lamprey in the Great Lakes. *Transactions of the American Fisheries Society* 99: 766-75.

Lawrie, A. H., and W. MacCallum. 1980. On evaluating measures to rehabilitate lake trout *(Salvelinus namaycush)* of Lake Superior. *Canadian Journal of Fisheries and Aquatic Sciences* 37: 2057-62.

Lawrie, A. H., and Jerold F. Rahrer. 1973. *Lake Superior: A case history of the lake and its fisheries.* Great Lakes Fishery Commission, Technical Report no. 19. Ann Arbor.

Loftus, K. H. 1958. Studies on river-spawning populations of lake trout in eastern Lake Superior. *Transactions of the American Fisheries Society* 87: 259-77.

MacDonald, Graham A. 1977. The ancient fishery at Sault Ste. Marie. *Canadian Geographical Journal* 94, no. 2 (April-May 1977): 54-58.

MacKay, H. H. 1963. *Fishes of Ontario.* Toronto: Ontario Department of Lands and Forests.

Manion, Patrick J., and Lee H. Hanson. 1980. Spawning behavior and fecundity of lampreys from the upper three Great Lakes. *Canadian Journal of Fisheries and Aquatic Sciences* 37: 1635-40.

Meyer, Fred P., and Rosalie A. Schnick. 1983. Sea lamprey control techniques: Past, present, and future. *Journal of Great Lakes Research* 9: 354-58.

Morman, R. H., D. W. Cuddy, and P. C. Rugen. 1980. Factors influencing the distribution of sea lamprey *(Petromyzon marinus)* in the Great Lakes. *Canadian Journal of Fisheries and Aquatic Sciences* 37: 1811-26.

Parsons, John W. 1973. *History of salmon in the Great Lakes, 1850-1970.* U. S. Bureau of Sport Fisheries and Wildlife, Technical Paper 68. Washington, D.C.

Pearce, W. A., R. A. Braem, S. M. Dustin, and J. J. Tibbles. 1980. Sea lamprey *(Petromyzon marinus)* in the lower Great Lakes. *Canadian Journal of Fisheries and Aquatic Sciences* 37: 1802-10.

Phillips, Gary L., William D. Schmid, and James C. Underhill. 1982. *Fishes of the Minnesota region.* Minneapolis: University of Minnesota Press.

Regier, Henry A. 1979. Changes in species composition of Great Lakes fish communities caused by man. *Transactions of the North American Wildlife and Natural Resources Conference* 44: 558-66.

Rostlund, Erhard. 1952. *Freshwater fish and fishing in native North America.* Berkeley and Los Angeles: University of California Press.

Schuldt, R. J., and R. Goold. 1980. Changes in the distribution of native lampreys in Lake

Superior tributaries in response to sea lamprey (*Petromyzon marinus*) control, 1953-1977. *Canadian Journal of Fisheries and Aquatic Sciences* 37: 1872-85.

Scott, W. B. 1957. Changes in the fish fauna of Ontario. In *Changes in the fauna of Ontario*, ed. F. A. Urquhart, pp. 19-25. Toronto: Royal Ontario Museum, University of Toronto Press.

Scott, W. B., and E. J. Crossman. 1973. *Freshwater fishes of Canada*. Fisheries Research Board of Canada, Bulletin 184. Ottawa.

Smith, Bernard R. 1971. Sea lampreys in the Great Lakes of North America. In *The biology of lampreys*, vol. 1, ed. M. W. Hardisty and I. C. Potter, pp. 207-47. London and New York: Academic Press.

Smith, Bernard R., and J. James Tibbles. 1980. Sea lamprey (*Petromyzon marinus*) in Lakes Huron, Michigan, and Superior: History of invasion and control, 1936-78. *Canadian Journal of Fisheries and Aquatic Sciences* 37: 1780-1801.

Smith, Bernard R., J. James Tibbles, and B. G. H. Johnson. 1974. *Control of the sea lamprey (Petromyzon marinus) in Lake Superior, 1953-70.* Great Lakes Fishery Commission, Technical Report no. 26. Ann Arbor.

Smith, Stanford H. 1968. Species succession and fishery exploitation in the Great Lakes. *Journal of the Fisheries Research Board of Canada* 25: 667-93.

Smith, Stanford H. 1970. Trends in fishery management of the Great Lakes. In *A century of fisheries in North America*, ed. Norman G. Benson, pp. 107-14. American Fisheries Society, Special Publication no. 7. Washington, D.C.

Smith, Stanford H. 1972. Factors of ecologic succession in oligotrophic fish communities of the Laurentian Great Lakes. *Journal of the Fisheries Research Board of Canada* 29: 717-30.

Talhelm, Daniel R., and Richard C. Bishop. 1980. Benefits and costs of sea lamprey (*Petromyzon marinus*) control in the Great Lakes: Some preliminary results. *Canadian Journal of Fisheries and Aquatic Sciences* 37: 2169-74.

Taylor, Robert Stanley. 1950. The historical development of the four Welland Canals, 1824-1933. M.A. thesis, University of Western Ontario, London, Ontario.

Torblaa, R. L., and R. W. Westman. 1980. Ecological impacts of lampricide treatments on sea lamprey (*Petromyzon marinus*) ammocoetes and metamorphosed individuals. *Canadian Journal of Fisheries and Aquatic Sciences* 37. 1835-50.

Underhill, James C. 1986. The fish fauna of the Laurentian Great Lakes, the St. Lawrence Lowlands, Newfoundland, and Labrador. In *The zoogeography of North American freshwater fishes*. ed. Charles H. Hocutt and E. O. Wiley, pp. 105-36. New York: John Wiley.

Underhill, James C., and John B. Moyle. 1968. The fishes of Minnesota's Lake Superior region. *The Conservation Volunteer* 32, no. 171 (January-February): 29-53.

Van Oosten, John. 1937. The dispersal of smelt, *Osmerus mordax* (Mitchell), in the Great Lakes region. *Transactions of the American Fisheries Society* 66: 160-71.

Walters, Carl J., George Spangler, W. J. Christie, Patrick J. Manion, and James F. Kitchell. 1980. A synthesis of knowns, unknowns, and policy recommendations from the Sea Lamprey International Symposium. *Canadian Journal of Fisheries and Aquatic Sciences* 37: 2202-8.

Walters, Carl J., Greg Steer, and George Spangler. 1980. Responses of lake trout (*Salvelinus namaycush*) to harvesting, stocking, and lamprey reduction. *Canadian Journal of Fisheries and Aquatic Sciences* 37: 2133-45.

Young, Gordon. 1973. Superior-Michigan-Huron-Erie-Ontario, is it too late? *National Geographic* 144: 147-85.

# Chapter 11. The Fish Market
# Chapter 12. Fishing for Fun

Adams, G. F., and D. P. Kolenosky. 1974. *Out of the water, Ontario's freshwater fish industry.* Toronto: Ontario Ministry of Natural Resources.

Armstrong, K. B., C. Hartviksen, B. J. Ritchie, and C. Vannatto. 1983. *The evaluation of the recreational fishery in Thunder Bay.* Toronto: Ontario Ministry of Natural Resources.

Baldwin, Norman S., Robert W. Saalfeld, Margaret A. Ross, and Howard J. Buettner. 1979. *Commercial fish production in the Great Lakes, 1867-1977.* Great Lakes Fishery Commission, Technical Report no. 3. Ann Arbor.

Bell, Howard. 1985. Comeback for Lake Superior's vanished herring? *Minnesota Volunteer* 48, no. 278 (January-February): 2-10.

Bolen, Don. 1978. *The ballad of Ragnvald Sve.* Two Harbors, Minn.: Don Bolen.

Christie, W. J. 1974. Changes in the fish species composition of the Great Lakes. *Journal of the Fisheries Research Board of Canada* 31: 827-54.

Close, Tracy L., Steven E. Colvin, and Richard L. Hassinger. 1984. *Chinook salmon in the Minnesota sport fishery of Lake Superior.* Minnesota Department of Natural Resources, Division of Fish and Wildlife, Section of Fisheries, Investigational Report no. 380. St. Paul.

Close, Tracy, and Richard Hassinger. 1981. *Evaluation of Madison, Donaldson, and Kamloops strains of rainbow trout* (Salmo gairdneri) *in Lake Superior.* Minnesota Department of Natural Resources, Division of Fish and Wildlife, Section of Fisheries, Investigational Report no. 372. St. Paul.

Gollat, R. 1976. *The fish and wildlife values of the Black Bay peninsula and offshore Lake Superior islands.* Thunder Bay: Ontario Ministry of Natural Resources.

Great Lakes Fishery Commission. 1983. *Annual report for the year 1981.* Ann Arbor.

Hassinger, R. L. 1974. *Evaluation of coho salmon* (Oncorhynchus kisutch) *as a sport fish in Minnesota.* Minnesota Department of Natural Resources, Division of Fish and Wildlife, Section of Fisheries, Investigational Report no. 328. St. Paul.

Hassinger, R. L. 1978. Fishing the great lake. *Minnesota Volunteer* 41, no. 239 (July-August): 20-27.

Hassinger, R. L., J. G. Hale, and D. E. Woods. 1974. *Steelhead of the Minnesota North Shore.* Minnesota Department of Natural Resources, Division of Fish and Wildlife, Section of Fisheries, Technical Bulletin no. 11. St. Paul.

Hile, Ralph, Paul H. Eschmeyer, and George F. Lunger. 1951. Status of the lake trout fishery in Lake Superior. *Transactions of the American Fisheries Society* 80: 278-312.

Holmquist, June Drenning. 1955. Commercial fishing on Lake Superior in the 1890s. *Minnesota History* 34: 243-49.

Johnson, Merle W. 1978. *The management of lakes for stream trout and salmon.* Minnesota Department of Natural Resources, Division of Fish and Wildlife, Section of Fisheries, Special Publication no. 125. St. Paul.

Kaups, Matti. 1978. North Shore commercial fishing, 1849-1870. *Minnesota History* 46: 43-58.

Kwain, Wen-hwa, and Andrew H. Lawrie. 1981. Pink salmon in the Great Lakes. *Fisheries* 6: 2-6.

Landin, Ed. 1983. *A Great Lakes fisherman.* Au Train, Mich.: Avery Color Studios.

Lawrie, A. H., and Jerold F. Rahrer. 1972. Lake Superior: Effects of exploitation and introductions on the salmonid community. *Journal of the Fisheries Research Board of Canada* 29: 765-76.

Lawrie, A. H., and Jerold F. Rahrer. 1973. *Lake Superior: A case history of the lake and its fisheries.* Great Lakes Fishery Commission, Technical Report no. 19. Ann Arbor.

MacCrimmon, Hugh R., and Barra Lowe Gots. 1972. *Rainbow trout in the Great Lakes.* Toronto: Ontario Ministry of Natural Resources, Sport Fisheries Branch.

MacDonald, Graham A. 1974. *East of Superior: A history of the Lake Superior Provincial Park region.* Ontario Ministry of Natural Resources, Parks Branch, Wawa District Office.

Martin, N. V., and C. H. Olver. 1976. *The distribution and characteristics of Ontario lake trout lakes.* Ontario Ministry of Natural Resources, Division of Fish and Wildlife, Research Report no. 97. Toronto.

Nute, Grace Lee. 1926. The American Fur Company's fishing enterprises on Lake Superior. *The Mississippi Valley Historical Review* 12: 483-503.

Ontario Ministry of Natural Resources. 1985. *Lake Superior strategic fisheries plan, 1985-2000, background information and optional management strategies.* Toronto.

Ontario Ministry of Natural Resources. 1986. Thunder Bay District fisheries management plan, 1986-2000. Toronto. (Draft.)

Peterson, Arthur R. 1971. *Fish and game lake resources in Minnesota.* Minnesota Department of Natural Resources, Division of Game and Fish, Section of Technical Services, Special Publication no. 89. St. Paul.

Phillips, Gary L., William D. Schmid, and James C. Underhill. 1982. *Fishes of the Minnesota region.* Minneapolis: University of Minnesota Press.

Rahrer, Jerold F. 1965. Age, growth, maturity, and fecundity of "humper" lake trout, Isle Royale, Lake Superior. *Transactions of the American Fisheries Society* 94: 75-83.

Scott, W. B., and E. J. Crossman. 1973. *Freshwater fishes of Canada.* Fisheries Research Board of Canada, Bulletin 184. Ottawa.

Selgeby, J. H., W. R. MacCallum, and D. V. Swedberg. 1978. Predation by rainbow smelt (*Osmerus mordax*) on lake herring (Coregonus artedii) in western Lake Superior. *Journal of the Fisheries Research Board of Canada* 35: 1457-63.

Smith, Lloyd L., Jr., and John B. Moyle. 1944. *A biological survey and fishery management plan for the streams of the Lake Superior North Shore watershed.* Minnesota Department of Conservation, Division of Game and Fish, Technical Bulletin no. 1. St. Paul.

Smith, Stanford H. 1968. Species succession and fishery exploitation in the Great Lakes. *Journal of the Fisheries Research Board of Canada* 25: 667-93.

Smith, Stanford H. 1972. Factors of ecologic succession in oligotrophic fish communities of the Laurentian Great Lakes. *Journal of the Fisheries Research Board of Canada* 29: 717-30.

Underhill, James C., and John B. Moyle. 1968. The fishes of Minnesota's Lake Superior region. *The Conservation Volunteer* 31, no. 171 (January-February): 1-25.

University of Minnesota Sea Grant. 1984. What happened to herring. *The Seiche* 8: 1-4.

Van Oosten, John. 1937. The dispersal of smelt, *Osmerus mordax* (Mitchill), in the Great Lakes region. *Transactions of the American Fisheries Society* 66: 160-71.

Weiler, John M. 1980. Michipicoten: Hudson's Bay Company post, 1821-1904. In *Three heritage studies,* ed. David Skene Melvin, pp. 4-64. Ontario Ministry of Culture and Recreation, Heritage Planning Study 5. Toronto.

Wyspianski, C. E. 1975. An examination of *Petromyzon marinus* L. (Sea Lamprey) and the effect of its invasion on the commercial fish community of northern Lake Superior, 1946-1970. Bachelor of Science in Geography-Biology, Dissertation, Lakehead University, Thunder Bay.

# Chapter 13. Wings Along the Coast

Bendel, J. F., and F. C. Zwickel. 1979. Problems in the abundance and distribution of the blue, spruce, and ruffed grouse in North America. In *Woodland Grouse Symposium, 1978,* ed. T. W. I. Lovel, pp. 48-63. Suffolk, United Kingdom: World Pheasant Association.

Bent, Arthur Cleveland. 1921. *Life histories of North American gulls and terns. Order Longipennes.* Smithsonian Institution, U.S. National Museum, Bulletin 113. Washington, D.C.

Bent, Arthur Cleveland. 1922. *Life histories of North American petrels and pelicans and their allies. Order Tubinaries and Order Steganopodes.* Smithsonian Institution, U.S. National Museum, Bulletin 121. Washington, D.C.

Bent, Arthur Cleveland. 1923. *Life histories of North American wild fowl. Order Anseres (Part).* Smithsonian Institution, U.S. National Museum, Bulletin 126: 1-30. Washington, D.C.

Bent, Arthur Cleveland. 1927. *Life histories of North American shorebirds: Order Limicolae (Part 1).* Smithsonian Institution, U.S. National Museum, Bulletin 142. Washington, D.C.

Bent, Arthur Cleveland. 1929. *Life histories of North American shore birds.* Smithsonian Institution, U.S. National Museum, Bulletin 146: 78-97. Washington, D.C.

Blokpoel, H., J. P. Ryder, I. Seddon, and W. R. Carswell. 1980. *Colonial waterbirds nesting in Canadian Lake Superior in 1978.* Canadian Wildlife Service, Progress Notes, no. 118. Ottawa.

Bohm, Robert T. 1982. The ring-bills come back. *Minnesota Volunteer* 45, no. 260 (January-February): 11-15.

Burger, Joanna, and Bori L. Olla, eds. 1984. *Shorebirds: Migration and foraging behavior.* Behavior of marine animals, Current Perspectives in Research, volume 6. New York and London: Plenum Press.

Dawson, J. Blair. 1972. *The ruffed grouse in Ontario.* Toronto: Ontario Ministry of Natural Resources, Division of Fish and Wildlife.

Eckert, Kim R. 1980. Avian spectacular at Hawk Ridge. *Minnesota Volunteer* 43, no. 251 (July-August): 46-53.

Erickson, Laura, and LeAne Rutherford. 1984. Peregrine gamble on the North Shore. *Minnesota Volunteer* 47, no. 277 (November-December): 14-19.

Godfrey, W. Earl. 1966. *The birds of Canada.* National Museum of Canada, Bulletin no. 203, Biological Series no. 73. Ottawa.

Goodermote, Don. 1980. Herring Gull nest counts on the North Shore of Lake Superior. *The Loon* 52, no. 1 (Spring): 15-17.

Graham, Frank, Jr. 1975. *Gulls: An ecological history.* New York: Van Nostrand Reinhold.

Green, Janet C. 1975. Rare sea ducks at Grand Marais. *The Loon* 47, no. 1 (Spring): 11-12.

Green, Janet C., and Robert B. Janssen. 1975. *Minnesota birds—where, when, and how many.* Minneapolis: University of Minnesota Press.

Green, Janet C., and Gerald J. Niemi. 1980. *Birds of the Superior National Forest.* Washington, D.C.: Superior National Forest, Forest Service, U.S. Department of Agriculture.

Green, John C. 1972. Geology of the Hawk Ridge Nature Reserve. Duluth: University of Minnesota. (Unpublished report.)

Gullion, Gordon. 1984. *Grouse of the North Shore.* Oshkosh, Wis.: Willow Creek Press.

Hofslund, P. B. 1959. Fall migration of Herring Gulls from Knife Island, Minnesota. *Bird-Banding* 30: 104-14.

Johnsgard, Paul A. 1973. *Grouse and quails of North America.* Lincoln: University of Nebraska.

Robinson, William L. 1969. Habitat selection by spruce grouse in northern Michigan. *Journal of Wildlife Management* 33: 113-20.

Robinson, William L. 1980. *Fool hen: The spruce grouse on the Yellow Dog Plains.* Madison: University of Wisconsin Press.

Robinson, William L., and Eric G. Bolen. 1984. *Wildlife ecology and management*. New York: Macmillan.

Rutherford, LeAne H., and Laura Erickson. 1985. The return of the wanderer, bringing the peregrine back to the North Shore. *Lake Superior Port Cities* 6, issue 4 (Spring): 44-49.

Schimpf, Ann. 1979. Hawk watch. *Lake Superior Port Cities* 3, issue 1: 39-41.

Stout, Gardner D. (ed.), Peter Matthiessen (text), Robert Verity Clem (paintings), and Ralph S. Palmer (species accounts). 1967. *The shorebirds of North America*. New York: Viking Press.

Taverner, P. A. 1943. *Birds of Canada*. Toronto: Musson Book Company.

# Chapter 14. The Antlered Legions

Aho, Robert W., and Peter A. Jordan. 1070. Production of aquatic macrophytes and its utilization by moose on Isle Royale National Park. Proceedings of the *First Conference on Scientific Research in the National Parks, U.S.D.I., National Park Service Transactions and Proceedings* series 5(1): 341-48.

Allin, Albert E. 1959. The North Shore's changing wildlife. *Naturalist* 10, no. 4: 24-29.

Anderson, Roy C. 1972. The ecological relationships of meningeal worm and native cervids in North America. *Journal of Wildlife Diseases* 8: 304-10.

Anderson, Roy C., and A. K. Prestwood. 1981. Lungworms. In *Diseases and parasites of white-tailed deer*, ed. William R. Davidson, Frank A. Hayes, Victor F. Nettles, and Forest E. Kellogg, pp. 266-317. Second International White-tailed Deer Disease Symposium, University of Georgia, Tall Timbers Research Station, Miscellaneous Publication no. 7. Atlanta.

Banfield, A. W. F. 1974. *The mammals of Canada*. National Museum of Natural Sciences, National Museums of Canada. Toronto: University Toronto Press.

Bergerud, A. T., W. Wyett, and B. Snider. 1983. The role of wolf predation in limiting a moose population. *Journal of Wildlife Management* 47: 977-88.

Berggren, Kim. 1983. Moose disease: Fifty-year mystery solved. *Minnesota Volunteer* 46, no. 271: 10-15.

Breckenridge, W. J. 1949. A century of Minnesota wild life. *Minnesota History* 30: 123-34.

Cringan, A. T. 1956. Some aspects of the biology of caribou and a study of the woodland caribou range of the Slate Islands, Lake Superior, Ontario. M.A. thesis, Department of Zoology, University of Toronto.

Cumming, H. G. 1972. *The moose in Ontario*. Toronto: Ontario Ministry of Natural Resources, Division of Fish and Wildlife.

de Vos, Antoon. 1964. Range changes of mammals in the Great Lakes region. *American Midland Naturalist* 71: 210-31.

de Vos, Antoon, and Randolph L. Peterson. 1951. A review of the status of woodland caribou (*Rangifer caribou*) in Ontario. *Journal of Mammalogy* 32: 329-37.

Euler, David. 1979. *Vegetation management for wildlife in Ontario*. Toronto: Ontario Ministry of Natural Resources.

Gollat, R. 1976. The fish and wildlife values of the Black Bay peninsula and offshore Lake Superior islands. Toronto: Ontario Ministry of Natural Resources. (Unpublished report.)

Hazard, Evan B. 1982. *The mammals of Minnesota*. Minneapolis: University Minnesota Press.

BIBLIOGRAPHY

Karns, Patrick D. (c. 1980). *Environmental analysis report, reintroduction of woodland caribou, Superior National Forest.* St. Paul: Minnesota Department of Natural Resources.

Kolenosky, George B., Dennis R. Voigt, and R. O. Standfield. n. d. *Wolves and coyotes in Ontario.* Toronto: Ontario Ministry of Natural Resources.

Mech, L. David, and L. D. Frenzel, Jr., eds. 1971. *Ecological studies of the timber wolf in northeastern Minnesota.* United States Forest Service, Research Paper NC-52. Washington, D.C.

Mech, L. David, and Patrick D. Karns. 1977. *Role of the wolf in a deer decline in the Superior National Forest.* United States Forest Service, Research Paper NC-148. Washington, D.C.

Moyle, John B. 1965. Border country wildlife . . . 150 years ago! *The Conservation Volunteer* 28, no. 162 (July-August): 52-58.

Peterson, Randolph L. 1955. *North American moose.* Toronto: University Toronto Press.

Peterson, Randolph L. 1957. Changes in the mammalian fauna of Ontario. In *Changes in the fauna of Ontario*, ed. F. A. Urquhart, pp. 43-58. Toronto: Royal Ontario Museum, University Toronto Press.

Peterson, Randolph L. 1966. *The mammals of eastern Canada.* Toronto: Oxford University Press.

Peterson, William J. 1981. Coming of the caribou. *Minnesota Volunteer* 44, no. 259 (November-December): 17-23.

Schmidt, John L., and Douglas L. Gilbert. 1978. *Big game of North America, ecology and management.* Harrisburg, Penn.: Stackpole Books.

Shiras, George, 3d. 1921. The wild life of Lake Superior, past and present. *National Geographic* 40: 113-204.

Smith, H. L., and P. L. Verkruysse. 1983. *The white-tailed deer in Ontario, its ecology and management.* Toronto: Ontario Ministry of Natural Resources.

Smith, Peter, and E. L. Borczon. 1977. Managing for deer and timber. *Your Forests* 10, no. 1 (Spring): 1-12.

Timmermann, H. R., J. B. Snider, and D. Euler. 1975. The Slate Islands. Toronto: Ontario Ministry of Natural Resources. (Unpublished report.)

Urquhart, F. A., ed. 1957. *Changes in the fauna of Ontario.* Toronto: Royal Ontario Museum, University Toronto Press.

# *Chapter 15. The Nipigon Basin*

Adamstone, F. B., and W. J. K. Harkness. 1923. *The bottom organisms of Lake Nipigon.* University of Toronto Studies, Publications of the Ontario Fisheries Research Laboratory, no. 15, pp. 121-70.

Bell, Robert. 1870. *Report on Lakes Superior and Nipigon.* Report of Progress, Geological Survey of Canada from 1866 to 1869, pp. 313-64. Ottawa.

Clemens, Wilbert A. 1923. *The limnology of Lake Nipigon.* University of Toronto Studies, Publications of the Ontario Fisheries Research Laboratory, no. 11, pp. 1-31.

Clemens, Wilbert A., John R.Dymond, N. K. Bigelow, F. B. Adamstone, and W. J. K. Harkness. 1923. *The food of Lake Nipigon fishes.* University of Toronto Studies, Publications of the Ontario Fisheries Research Laboratory, no. 16, pp. 171-88.

Coleman, A. P. 1922. *Glacial and pre-glacial lakes in Ontario.* University of Toronto Studies, Publications of the Ontario Fisheries Research Laboratory, no. 10.

Cook, Marylyn. 1973. *Man on Lake Nipigon.* Toronto: Ontario Ministry of Natural Resources.

Dawson, K. C. A. 1967. *A report on archaeological reconnaissance of Lake Nipigon.* Toronto: Ontario Public Archives.

Dawson, K. C. A. 1970. *Archaeological investigations at the site of the trading post at Sand Point, Blacksand Provincial Park, Ontario.* Toronto: Ontario Department of Lands and Forests, Parks Branch.

Dawson, K. C. A. 1976. *Algonkians of Lake Nipigon.* National Museum of Man, Mercury Series, Archeological Survey of Canada, Paper no. 48. Ottawa.

Dymond, J. R. 1923. *A provisional list of the fishes of Lake Nipigon.* University of Toronto Studies, Publications of the Ontario Fisheries Research Laboratory, no. 12, pp. 33-38.

Dymond, J. R., L. L. Snyder, and E. B. S. Logier. 1928. A faunal investigation of the Lake Nipigon region, Ontario. *Transactions of the Royal Canadian Institute,* 16, part 2: 233-77.

Gapen, Dan D. 1983. Muddler. *Trout.* 24, no. 1 (Winter): 34-41.

Gollat, R. L. 1975. *A preliminary investigation into the fish and wildlife values of the islands of Lake Nipigon.* Toronto: Ontario Ministry of Natural Resources.

Griffith, Ray. 1972. *The history of Lake Nipigon.* Toronto: Ontario Ministry of Natural Resources.

Kor, Philip S. G. 1978. *A brief geology and geomorphology of Lake Nipigon Provincial Park.* Toronto: Ontario Ministry of Natural Resources, Division of Outdoor Recreation, North Central Division.

McInnes, William. 1894. *Ontario, survey of Lake Nipigon.* Annual Report, Geological Survey of Canada, Volume 7, part A. Ottawa.

Ontario Ministry of Natural Resources. 1981. *Nipigon land use plan, background information.* Toronto.

Ontario Ministry of Natural Resources. 1983. *Nipigon District land use guidelines.* Toronto.

Scott, Nancy. 1983. *Lake Nipigon Provincial Park, background information.* Toronto: Ontario Ministry of Natural Resources.

Teller, James T., and L. Harvey Thorleifson. 1983. The Lake Agassiz-Lake Superior connection. In *Glacial Lake Agassiz,* ed. J. T. Teller and Lee Clayton, pp. 261-90. Geological Association of Canada, Special Paper 26. Ottawa.

Wilson, Alfred W. G. 1910. *Geology of the Nipigon basin, Ontario.* Canada Department of Mines, Geological Survey Branch, Memoir no. 1. Ottawa.

# Chapter 16. The Royal Island

Aho, Robert W., and Peter A. Jordan. 1979. Production of aquatic macrophytes and its utilization by moose on Isle Royale National Park. *Proceedings of the First Conference on Scientific Research in the National Parks, U.S. National Park Service, Transactions and Proceedings* series 5: 341-48.

Aldous, Shaler E., and Laurits W. Krefting. 1946. The present status of moose on Isle Royale. *Transactions of the 11th North American Wildlife Conference:* 296-308.

Allen, Durward L. 1979. *Wolves of Minong.* Boston: Houghton Mifflin.

Anonymous. 1967. A wilderness plan for Isle Royale National Park and the surrounding region. *National Parks Magazine* 41: 18-19.

Bachmann, Elizabeth M. 1959. Isle Royale. *Naturalist* 10: 30-32.

Cooper, William S. 1913. The climax forest of Isle Royale, Lake Superior, and its development. I. *Botanical Gazette* 55: 1-44.

Eliot, John L. 1985. Isle Royale—a north woods park primeval. *National Geographic* 167: 534-50.

Foster, J. W., and J. D. Whitney. 1850. *Report on the geology and topography of a portion of the Lake Superior Land District in the State of Michigan. Part I. Copper lands.* United States 31st Congress, 1st Session, House Executive Document 69.

Fry, George W. 1961. Isle Royale. *Michigan Conservation* 30: 8-11.

# BIBLIOGRAPHY

Hansen, Henry L., Laurits W. Krefting, and Vilis Kurmis. 1973. *The forest of Isle Royale in relation to fire history and wildlife.* Minnesota Agricultural Experiment Station, Forestry Series 13, Technical Bulletin 294. St. Paul.

Holden, Thomas. 1979. The America: A friend to North Shore ports. *Lake Superior Port Cities* Spring: 30-32, 37.

Holte, Ingeborg. 1984. *Ingeborg's Isle Royale.* Grand Marais, Minn.: Women's Times Publishing.

Huber, N. King. 1975. *The geologic story of Isle Royale National Park.* United States Department of the Interior, National Park Service, Geological Survey Bulletin 1309. Washington, D.C.

Jordan, Peter A. 1981. *Wildlife of Isle Royale.* Houghton, Mich.: Isle Royale Natural History Association.

Jordan, Peter A., and Michael L. Wolfe. 1980. Aerial and pellet-count inventory of moose at Isle Royale. *Proceedings of the Second Conference on Scientific Research in the National Parks, United States National Park Service, Transactions and Proceedings* series 5: 363-93.

Karns, Patrick D., and Peter A. Jordan. 1969. *Pneumostrongylus tenuis* in moose on a deer-free island. *Journal of Wildlife Management* 33: 431-33.

Krefting, Laurits W. 1963. Beaver of Isle Royale. *Naturalist* 14: 2-11.

Krefting, Laurits W. 1969. The rise and fall of the coyote on Isle Royale. *Naturalist* 20: 24-31.

Krefting, Laurits W. 1974. *The ecology of the Isle Royale moose.* University of Minnesota, Agricultural Experiment Station, Forestry Series 15, Technical Bulletin 297. St. Paul.

Lagler, Karl F., and Charles R. Goldman. 1982. *Fishes of Isle Royale.* Houghton, Mich.: Isle Royale Natural History Association.

Linn, Robert M. 1962. *Forests and trees of Isle Royale National Park.* Houghton, Mich.: Isle Royale Natural History Association.

Mech, L. David. 1966. *The wolves of Isle Royale.* Fauna of the National Parks of the United States, Fauna Series 7. Washington, D.C.

Mech, L. David. 1970. *The wolf.* Garden City, N. Y.: Doubleday.

Murie, Adolph. 1934. *The moose of Isle Royale.* University of Michigan Museum of Zoology, Miscellaneous Publication no. 25. Ann Arbor.

Oikarinen, Peter. 1979. *Island folk, the people of Isle Royale.* Houghton, Mich.: Isle Royale Natural History Association.

Peterson, Rolf Olin. 1977. *Wolf ecology and prey relationships on Isle Royale.* United States National Park Service, Scientific Monograph Series, no. 11. Washington, D.C.

Rakestraw, Lawrence. 1965. *Historic mining on Isle Royale.* Houghton, Mich.: Isle Royale Natural History Association.

Rakestraw, Lawrence. 1968. *Commercial fishing on Isle Royale, 1800-1967.* Houghton, Mich.: Isle Royale Natural History Association.

Russell, Ann Z., and Yvonne Nissen. 1985-1986. The elusive *Algoma* bow. *Lake Superior Port Cities* 8, issue 1 (Winter 1985-86): 11-15.

Shelton, Napier. 1975. *The life of Isle Royale.* United States National Park Service, Natural History Series. Washington, D.C.

Stewart, Scot. 1982. Wolf music. *Michigan Natural Resources* November-December: 40-49.

Stonehouse, Frederick. 1983. *Isle Royale shipwrecks.* AuTrain, Mich.: Avery Color Studios.

Stucker, Gilbert F. 1969. Lake Superior's island wilderness. *National Parks* 43: 4-9.

# Chapter 17. A Heritage in Parks

## (see also Guides and Maps)

Campbell, George. 1985. Ouimet Canyon harbors an errant tundra. *Lake Superior Port Cities*, 7, issue 1 (Summer): 57-61.

deLaittre, John. (ca. 1980). *Tettegouche*. Minneapolis: privately published.

Eliseuson, Michael. 1976. *Tower Soudan . . . the state park down under.* Minnesota Parks Foundation, Minnesota State Park Heritage Series no. 1. St. Paul.

Fritzen, John. 1974. *Historic sites and place names of Minnesota's North Shore.* Duluth: St. Louis County Historical Society.

Fritzen, John. 1978. *The history of Fond du Lac and Jay Cooke Park.* Duluth: St. Louis County Historical Society.

Furtman, Michael. 1984. Crosby-Manitou State Park. *Lake Superior Port Cities* 6, issue 1: 58-60.

Grinstead, David L. 1986. Tettegouche . . . jewel of the state parks. *Lake Superior Port Cities* 8, issue 3 (May-June): 30-35.

Hall, Stephen P. 1978. *Split Rock: Epoch of a lighthouse.* Minnesota Historic Sites Pamphlet Series no. 15. St. Paul: Minnesota Historical Society.

Knopp, Timothy B., and Uel Blank. 1983. *The North Shore experience.* Minnesota Sea Grant Program, University of Minnesota, Research Report no. 8. Duluth.

MacDonald, Graham A. 1974. *East of Superior: A history of the Lake Superior Provincial Park region.* Toronto: Ontario Ministry of Natural Resources, Parks Branch.

Marsh, John S. 1976. *The human history of the Pukaskwa National Park area, 1650 to 1975, an initial study.* Ottawa: Parks Canada, Pukaskwa National Park.

Mountain, James A. 1974. *The inhospitable shore: An historical resource study of Neys Provincial Park.* Toronto: Ontario Ministry of Natural Resources, Division of Parks, Historical Sites Branch.

Ontario Ministry of Natural Resources. 1977. *Neys Provincial Park, master plan.* Toronto.

Ontario Ministry of Natural Resources. 1979. *Lake Superior Provincial Park, master plan.* Toronto.

Parks Canada. 1982. *Pukaskwa National Park, management plan.* Ottawa.: Parks Canada, Pukaskwa National Park.

Peake, Geoffrey. 1985. Pukaskwa: The best way to explore it is by canoe. *Canadian Geographic* 105, no. 1 (February-March): 30-36.

Revill, A. D., associates. 1980. *Oral history of Pukaskwa National Park.* Ottawa: Parks Canada, Pukaskwa National Park.

Sandvik, Glenn. 1972. *A superior beacon, a brief history of Split Rock Lighthouse.* Privately published.

Searle, R. Newell. 1979. *State parks of the North Shore.* Minnesota State Park Heritage Series no. 3. St. Paul: Minnesota Parks Foundation.

Vosper, G. F. 1984a. *The development of Lake Superior Provincial Park.* Toronto: Ontario Ministry of Natural Resources, Wawa District.

Vosper, G. F. 1984b. *Tourism in Lake Superior Provincial Park.* Toronto: Ontario Ministry of Natural Resources, Wawa District.

# Index

(Boldfaced page numbers refer to maps.)

# INDEX

Animikie sea, 14, 89
Anjigami Lake, **295**
Applegate, Vernon C., fishery scientist, 154
Archaeological remains in Lake Superior region: prehistoric, 30-35; European, at Michipicoten, 76-77; at Lake Nipigon, 248, 300
Argo, Minn. *See* Babbitt, Minn.
Army Corps of Engineers, U.S., 309
Artists Point, Grand Marais, Minn., 303
Ashburton, Lord, British envoy, 82
Ashland, Wis., **111**, 112, 114
Aspen, 108, 124, 125, 133, 267, 277
Assiniboin Indians, 35: distribution of, **36**
Astor, John Jacob, American fur trader, 71, 76
Athabasca, Lake, 57, 67: lake trout in, 203
Audubon Societies, 223
Aurora, Minn., **96**, 100, 101
Austin, James M., lumber mill owner, 122-23
Austin Nicholson Company, 121-23, 290

Babbitt, Minn., **96**, 100, 102, 103
Backus Brooks Enterprises, forest products company, 133
Baldhead River, **275**
Baptism River, **209**, 301: rainbow trout in, 208
Baptism River State Park. *See* Tettegouche State Park
Baraga, Mich., **111**, **127**: lumber mills at, 112, 114
Barrow's Goldeneye, 224
Bass, smallmouth, 210-11
Batchawana Bay, Lake Superior, 24, 186, **187**
Batchawana Provincial Park, **285**, 291
Batchawana River, 125, **127**
Bayer 73, molluscide used in sea lamprey control, 154
Bear, black, 32: in Nipigon Basin, 256; not on Isle Royale, 278
Beaver: as food for prehistoric humans, 32; in fur trade, 39, 42, 60, 66, 68, 73, 75; decline of, 77-80, 253; as food for gray wolf, 241; on Isle Royale, 267-68, 277; in Neys Provincial Park, 299
Beaver Bay, Minn., **x**, **96**, 97, 177, 179: log landings on, 111
Beaver River, **167**, 294

Bensen Lake, 301
Bering Land Bridge, 26-27, 30, 35, 232
Beringia. *See* Bering Land Bridge
Big Gravel River, treated for sea lampreys, 155
Birch, white, 108, 124, 133, 267
Birch Lake, **96**, 97, 98, 100
Birchbark canoe, 5, 41, 62-63, 64, 65, 67-68, 71, 74, 184
Bison, giant, 26
Biwabik, Minn., **96**, 100
Black Bay, Lake Superior, **x**, **88**, **127**, 133, 167, 187, 209, 221, 246, 249, 285: herring fishery in, 191; brook trout in, 200
Black Bay Peninsula, 17, **88**, **127**, 133, 199, 291
Black River, **127**, 130, 131
Black Sturgeon Bay, Lake Nipigon, **249**
Black Sturgeon Lake, **249**
Black Sturgeon River, **127**, 132, 246, 249
Blacksand Provincial Park. *See* Lake Nipigon Provincial Park
Blister rust, disease of white pine, 109, 134
Bloater, 142, 162, 169, 192
Bog lakes, managed for trout, 199
*Bon Ami*, steam tug used in commercial fishing, 179
*Bonasa umbellus. See* Grouse, Ruffed
Booth Fisheries Company, Chicago, 188
Booth Packing Company, A., fish company at Duluth, 179
Boreal Forest Region, 37, **111**, 119-35 *passim*, **127**, **221**, 240, 313; character of, 7, 106; succession of, 28-30, 32; distribution of, 106, 116-17, 119; bird fauna in, 227-28; prehistoric Indians in, 250; in Nipigon Basin, 259; in Lake Superior Provincial Park, 298; in Sibley Provincial Park, 300
Boundary Waters Canoe Area, 241
Boyer, Ben, discoverer of Helen Iron Mine deposits, 92-93
Boyer Lake, 93
Brandsford Reef, Isle Royale, **264**, 279
*Bransford*, steamship, 279
Bray Fish Company, Duluth, 179
Brighton Beach, Duluth, 302-3
Brooks-Scanlon Lumber Company, 114
Brookston, Minn., 99
Brulé, Étienne, explorer, 41, 261, 308

344

# INDEX

# INDEX

**Thomas F. Waters** has been a professor in the Department of Fisheries and Wildlife at the University of Minnesota since the late 1950s. For more than 25 years he has studied the streams and rivers of Minnesota, as a limnologist, professor, conservationist and trout fisherman. Waters received his B.S. and M.S. in zoology with an emphasis on aquatics, and his Ph.D. in fisheries and wildlife from Michigan State University. His family owned a log cabin on the shore of Speckled Trout Lake, in the district of Algoma, Ontario, and as a teenager, Waters went with his father on fishing trips in search of brook trout. On Speckled Trout Lake, Waters's interest in the Lake Superior region first developed. He has written numerous scientific papers on stream ecology; he is the author of *The Streams and Rivers of Minnesota*, (Minnesota, 1977) and contributes to such journals as *Ecology, Limnology and Oceanography, Canadian Journal of Fisheries and Aquatic Sciences*, and *Transactions of the American Fisheries Society*.

**Carol Yonker Waters** is an artist whose paintings and drawings have been exhibited in the Twin Cities area; she has served as president of the board of directors for Dakota Center for the Arts. Her family owned a cabin in northern Michigan, where she learned to appreciate woods and streams. Tom and Carol Waters's research for this book included a year's travel together on the North Shore. "Leaning on a warm Lake Superior rock and getting lost in drawing is terrific," she says, "but it is the area itself that quickens the spirit, and anyone can experience that."